Table of Contents

Page

List of Figures iv

List of Tables vi

List of Symbols vii

I. Introduction 1

    1.1   Motivation 1
    1.2   Problem Statement 3
        1.2.1   Case 1: Impulsive In-Plane Thrusting 4
        1.2.2   Case 2: Impulsive Out-Of-Plane Thrusting 5
        1.2.3   Case 3: Continuous Thrusting 5
    1.3   Method of Investigation 6
    1.4   Thesis Overview 6

II. Background 7

    2.1   Chapter Overview 7
    2.2   Related Work 7
    2.3   Equations of Motion 9
        2.3.1   The Two-Body Problem 9
        2.3.2   Classical Orbital Elements 12
        2.3.3   Equinoctial Orbital Elements 18
    2.4   Optimal Control Theory 23
        2.4.1   The General Indirect Method 23
        2.4.2   Primer Vector Theory 26
        2.4.3   Pseudospectral Methods 29
    2.5   Chapter Summary 32

III. Methodology 33

    3.1   Chapter Overview 33
    3.2   Optimal Control Problem Formulation 33

		Page

    3.2.1   Equations of Motion 33

    3.2.2   State and Control Constraints 35

    3.2.3   Cost Function 38

    3.2.4   Multiple vs Single Phase Problem 39

  3.3   General Pseudospectral Optimal Control Software-II 41

    3.3.1   Overview 41

    3.3.2   Input Structure 42

    3.3.3   Additional Required Functions 43

    3.3.4   Output Structure 44

    3.3.5   Limitations 44

  3.4   Systems Tool Kit® v 10 45

    3.4.1   Component Object Model Interface Library 45

    3.4.2   Scenario Input 46

    3.4.3   Maneuver Development 49

    3.4.4   Optimizer Result Validation 51

  3.5   Chapter Summary 52

IV. Analysis and Results 53

  4.1   Chapter Overview 53

  4.2   Optimal Control Results 53

    4.2.1   Case 1 Single Orbit 54

    4.2.2   Case 1 Multiple Orbit 57

    4.2.3   Case 2 Single Orbit 59

    4.2.4   Case 2 Multiple Orbit 62

    4.2.5   Case 3 Single Orbit 62

    4.2.6   Case 3 Multiple Orbit 64

    4.2.7   Summary of Optimal Control Results 66

  4.3   Systems Tool Kit® Simulation and Validation 68

    4.3.1   Case 1 68

    4.3.2   Case 2 73

    4.3.3   Case 3 76

    4.3.4   Summary of STK Results 78

  4.4   Chapter Summary 81

V. Conclusions and Recommendations 82

		Page
5.1	Chapter Overview	82
5.2	Conclusions	83
5.3	Research Limitations	84
5.4	Recommendations for Future Work	84

Appendix A. MATLAB® Code 87

Appendix B. GPOPS-II Structure Architecture 109

Bibliography 119

List of Figures

                                                                                                       Page

Figure 1: Orbital Debris Population Growth [1] 2

Figure 2: Two-Body Problem [4] 11

Figure 3: Classical Orbital Elements [4] 13

Figure 4: Eccentric Anomaly 17

Figure 5: Equinoctial Reference Frame [16] 20

Figure 6: LG, LGR, and LGL collocation points [22] 31

Figure 7: Defect Vector [20] 32

Figure 8: Thrust Vector 34

Figure 9: Penalty Functions 40

Figure 10: Area of Regard 48

Figure 11: Error Ellipse 49

Figure 12: Optimization Routine Flowchart 50

Figure 13: Case 1 Single Orbit Thrusting Profile 55

Figure 14: Case 1 Single Orbit COE 55

Figure 15: Case 1 Single Orbit Relative Motion Cross Section 56

Figure 16: Case 1 Multiple Orbit Thrusting Profile 58

Figure 17: Case 1 Multiple Orbit COE 58

Figure 18: Case 1 Multiple Orbit Relative Motion Cross Section 59

Figure 19: Case 2 Thrusting Profile 60

Figure 20: Case 2 COE 60

Page

Figure 21: Case 2 Orbit Relative Motion Cross Section 61

Figure 22: Case 3 Single Orbit Thrusting Profile 63

Figure 23: Case 3 Single Orbit COE 63

Figure 24: Case 3 Single Orbit Relative Motion Cross Section 64

Figure 25: Case 3 Multiple Orbit Thrusting Profile 65

Figure 26: Case 3 Multiple Orbit COE 65

Figure 27: Case 3 Multiple Orbit Relative Motion Cross Section 66

Figure 28: Case 1 Single Orbit STK Results 70

Figure 29: Case 1 Single Orbit Pitch and Yaw Validation 70

Figure 30: Case 1 Single Orbit Thrust Duration Validation 71

Figure 31: Case 1 Multiple Orbit STK Results 72

Figure 32: Case 1 Multiple Orbit Pitch and Yaw Validation 72

Figure 33: Case 1 Multiple Orbit Thrust Duration Validation 73

Figure 34: Case 2 STK Results 74

Figure 35: Case 2 Pitch and Yaw Validation 75

Figure 36: Case 2 Thrust Duration Validation 75

Figure 37: Case 3 Single Orbit STK Results 76

Figure 38: Case 3 Single Orbit Pitch Profile Validation 77

Figure 39: Case 3 Multiple Orbit STK Results 78

Figure 40: Case 3 Multiple Orbit Pitch and Yaw Validation 79

Figure 41: Case 3 Multiple Orbit Thrust Duration Validation 79

List of Tables

Page

Table 1: Eccentricity 15

Table 2: Global State Constraints 36

Table 3: Control Constraints 37

Table 4: STK COM Interface Library Function List 47

Table 5: Area Target Parameters 48

Table 6: Satellite Initial States 49

Table 7: Fuel Cost Comparison 68

List of Symbols

Roman

A	Perturbing acceleration vector
a	Semi-major axis
E	Eccentric anomaly
e	Eccentricity
F	Eccentric longitude
[f, g, w]	Equinoctial reference frame unit vectors
G	Universal gravitational constant
$\mathcal{H}$	Hamiltonian
h	Eccentricity vector component in the *f* direction
I	Identity matrix
i	Inclination
[i, j, k]	Earth Centered Inertial reference frame unit vectors
J	Cost function
k	Eccentricity vector component in the *g* direction
M	Mean anomaly
m	Mass
n	Vector designating the Line of Nodes
n	Mean motion
P_N	Legendre Polynomial
p	Equinoctial element
q	Equinoctial element
r	Inertial position vector
r	Magnitude of the inertial position vector
T	Thrust magnitude
t	Time
t_0	Epoch time
u	Control vector
v	Inertial velocity vector
v	Magnitude of the inertial velocity vector
x	State vector
x	Position along the *i* vector

x_1	Position along the ***f*** vector
y	Position along the ***j*** vector
y_1	Position along the ***g*** vector
z	Position along the ***k*** vector

Greek

α	Cost function weighting factor
Γ	Thrust magnitude
$\boldsymbol{\Delta}$	Defect Vector
ε	Total mechanical energy
θ	Pitch angle
$\boldsymbol{\lambda}$	Vector of Lagrange multipliers
λ	Mean longitude
μ	System specific gravitational constant
ν	True anomaly
ψ	Yaw angle
Ω	Right ascension of the ascending node
ω	Argument of perigee

TRAJECTORY OPTIMIZATION FOR SPACECRAFT COLLISION AVOIDANCE

I. Introduction

1.1 Motivation

A half century of space research and development has left the near-earth environment littered with large quantities of orbital debris. Spent rocket bodies and dead satellites constitute the largest pieces of debris currently being tracked by the US Space Surveillance Network (SSN). However, in addition to these large pieces of debris there also exists a large volume of smaller objects formed from collisions between the larger debris. The estimate for total population in the near-earth environment as of April 2011 was 28,000 objects larger than 10 cm [1]. While the radar cross sections of the smaller debris makes tracking and cataloging more difficult, hundreds of thousands of objects are assumed to be in orbit on the 1 cm level and hundreds of millions of objects are expected at the 1 mm level [1]. Figure 1 details the estimated population growth of orbital debris over the past five decades. Liou [1] projected through the use of 100 Monte Carlo simulations the estimated growth over the next century. The 1-σ values for these projections are also included in this figure.

This trend has been a source of major concern to the international community for decades, prompting cooperative attempts to minimize this growth and preserve the accessibility of the near-earth environment. However, recent studies have shown that thus far the international efforts to mitigate the growth of orbital debris have not proven effective enough and the population of orbital debris continues to grow. Exploration on how to conduct active debris removal has also been a subject of research in recent years. Several proposals have been made such as a ground-based or space-based laser system or attaching inflatable balloons or sails to the larger debris to increase drag and decay the orbit. Thus far, however, no viable solution has been implemented to actively remove debris in orbit [1].

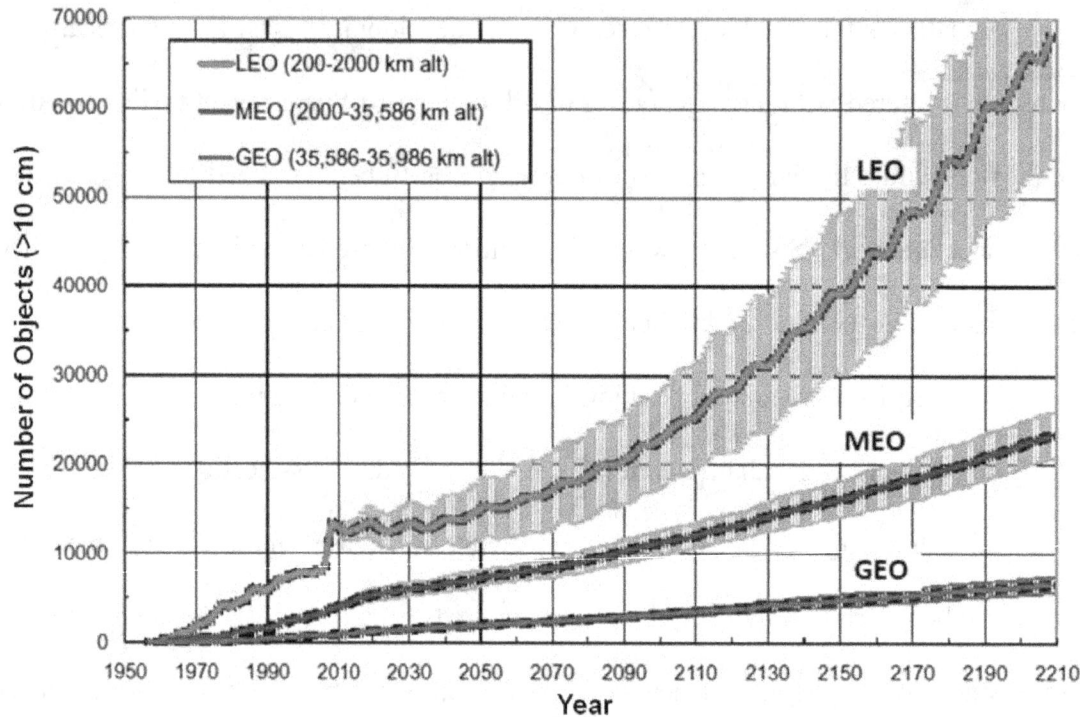

Figure 1: Orbital Debris Population Growth [1]

The space environment is divided into three orbital zones. The altitude band between 200 km and 2000 km is referred to as Low Earth Orbit (LEO) [1]. LEO has seen the largest volume of traffic of active satellites due to its relative accessibility as well as allowing for high signal strength communications with ground stations. The Geosynchronous (GEO) region spans the space within 200 km of the geosynchronous altitude of 36,000 km. This region is heavily populated by larger satellites in the Geostationary Arc which is located in the vicinity of the equator. In between these regions is defined as Medium Earth Orbit (MEO) and is primarily used by navigation satellites. While debris population growth is observed in all three of these regions, the vast majority of observed growth occurs in LEO [1]. Despite the fact that the debris in LEO tends to decay relatively quickly due to air drag, new debris is continually being introduced as smaller debris is formed from collisions in higher orbits that continually decay into this region. Recent observations have shown that the rate of decay of debris into LEO is nearly the same as the rate of decay of debris departing LEO but is expected to begin significantly increasing within the next 50 years due to expected collisions in MEO [2].

The average impact speed for a satellite in LEO is around 10 km/sec [3]. Even for collisions of satellites with objects as small as 5 mm, a hypervelocity impact has the potential to end a satellite's service life. As the volume of orbital debris increases, satellites are required to make an ever increasing number of maneuvers to avoid damage. The year 2010 alone saw nearly 400 warnings and over 100 avoidance maneuvers conducted in order to minimize the risk of collisions [1]. With each passing year, the number of necessary maneuvers increases with the population of debris in orbit. Each of these maneuvers detracts from the overall service life of the maneuvering satellite.

1.2 Problem Statement

The purpose of this research was to develop and test the application of pseudospectral optimization to orbital maneuvering. This was accomplished through the determination of a set of minimum fuel thrust profiles to maneuver a satellite for the purpose of collision avoidance. An Area of Regard (AOR) was specified in order to designate a 'no-thrust' region for this scenario. This region was necessary since thrusting degrades the performance of satellite payloads as well as complicating orbit determination solutions. Therefore, the start time for this scenario occurs when the satellite departs the AOR and terminates upon AOR reentry. Upon reentry, the maneuvering satellite is required to be outside a user-specified error ellipse projected from its non-thrusting reference trajectory. This research utilized an error ellipse that is 100 km in-track, 10 km out-of-track, and 10 km out-of-plane in size. This study looked at three distinct cases: Impulsive In-Plane, Impulsive Out-of-Plane, and Continuous In-Plane.

1.2.1 Case 1: Impulsive In-Plane Thrusting

Impulsive thrusting is the traditional method used to maneuver satellites. This method is relatively simple to model and provides large accelerations and a rapid satellite response to commanded maneuvers. It is capable of achieving nearly instantaneous velocity changes necessary for large orbital maneuvers. However, impulsive thrusting typically makes use of engines that have relatively low specific impulse (I_{SP}) and are therefore expensive to operate. Thrusting in the satellite's orbital plane is considered to be the least expensive maneuver and is therefore the first case considered. Conventional wisdom states that the minimum fuel thrust direction is either in the velocity or anti-velocity direction depending on whether a climb or a descent is desired. Due to a desire to keep the orbit circular, an impulsive thrust is typically conducted twice, once to climb or descend and once to re-circularize the orbit at the desired altitude. This maneuver is referred to as a Hohmann Transfer [4] and is most commonly used when an altitude change is desired. However, if the satellite mission permits small variances in eccentricity, it may not be optimal to recircularize the orbit after conducting a collision avoidance maneuver. For the Impulsive In-Plane thrusting case, this research demonstrates a more fuel efficient maneuver than the Hohmann Transfer for the purpose of collision avoidance.

1.2.2 Case 2: Impulsive Out-Of-Plane Thrusting

While thrusting out of the satellite's orbital plane is considered to be less fuel efficient than the previous case, it allows for alterations to the orbital plane itself and can therefore be a useful alternative method in collision avoidance. This case is less generalized than the previous case, however, since it depends heavily on the latitude of the AOR and the inclination of the orbital plane. Therefore, this research generates an algorithm to determine the optimal thrust time and direction for an unspecified set of latitudes and inclinations.

1.2.3 Case 3: Continuous Thrusting

Continuous thrust maneuvers utilize Electric Propulsion (EP) thrusters in order to generate their accelerations. They are used less often due to the extremely low forces they generate. However, these engines are more fuel efficient due to their extremely high I_{SP}. For this reason, the use of continuous thrust engines can therefore extend the service life of a maneuvering satellite that would otherwise rely on impulsive thrust engines for collision avoidance. This research demonstrates a method for maneuvering using continuous thrust that is comparable to the first case by utilizing thrust direction rather than duration.

1.3 Method of Investigation

The scenario start and termination times were developed using Analytical Graphics Incorporated® (AGI) Systems Tool Kit® (STK) version 10 via an access report generated between a non-maneuvering reference satellite and an AOR. The scenario start time along with the current state were then imported into MATLAB® for optimization using General Pseudospectral Optimal Control Software (GPOPS-II). This software utilized the Radau Pseudospectral Method (RPM) to optimize thrust/angle profiles for each of the three scenarios mentioned in the previous section. These profiles were then converted into a form that was accessible to STK. The Astrogator propagation tool was used in STK to test these thrust profiles and measure the distance at scenario termination from the reference satellite to a satellite with identical initial conditions executing each of the calculated optimal thrust profiles.

1.4 Thesis Overview

Chapter II provides the mathematical background required in order to formulate the necessary components of the Optimal Control Problem. It outlines several choices of states and their corresponding equations of motion as well as the general principles of Optimal Control Theory. Chapter III details the methods used in setting up the problem in GPOPS-II as well as STK. Chapter IV presents and discusses the results from the algorithm developed in Chapter III. Chapter V summarizes the conclusions from this research and presents suggestions for future study.

II. Background

2.1 Chapter Overview

This chapter establishes the basis for the methods used to determine the optimal orbital trajectories discussed in this research. First, a general discussion is made on the recent work this research is based on. Second, a derivation of the equations of motion is discussed. The benefits and difficulties inherent in several different choices of states are discussed as well as their corresponding equations of motion. Finally a discussion is presented on the background of the optimization methods used in later chapters.

2.2 Related Work

This research combines elements from previous work accomplished in the fields of responsive spacecraft control and optimal control techniques. The work from responsive spacecraft control formed the baseline for the formulation of the equations of motion as well as the choices of the three maneuver cases outlined in Section 1.2. The optimal control research cited in this section formed the baseline for the development of the algorithms used to optimize the trajectories presented in Chapter IV.

This research is most closely based on the work done by Co [5] and Zagaris [6]. In his 2012 dissertation, Co [5] explored the differences between electric and chemical propulsion and their applications in generating a desired change in the satellite's overflight time of a ground target. His work with electric propulsion along with the work accomplished by Zagaris [6] in his thesis formed the basis for the formulation of the equations of motion as well as the optimal control approach used in this research.

Zagaris utilized both optimal control methods as well as a Lyapunov control technique in order to modify the time of passage of a satellite over a specified ground target.

Jorris [7] and Karasz [8] utilized pseudospectral optimization in the derivation of an optimal trajectory for an autonomous reentry vehicle subject to specified 'no-fly zone' path constraints. In his 2007 dissertation, Jorris [7] utilized a direct collocation method to design a multiple-phase algorithm that optimized a three-dimensional trajectory subject to his specified no-fly constraints. Karasz [8] built on this research and demonstrated through a sensitivity analysis how changes in the locations of the 'no-fly zones' affected the solution. Yaple [9] also followed this research in the development of a more general trajectory optimization tool.

Darby [10] demonstrated the application of *hp*-adaptive pseudospectral methods in spacecraft maneuver optimization. He utilized this technique to determine maneuvering cost for a spacecraft in LEO executing orbital inclination changes with assistance from atmospheric forces. This was conducted using three impulsive maneuvers: one to de-orbit in order to conduct atmospheric dipping, a boost maneuver to direct the satellite to its final altitude, and a final re-circularizing maneuver. He concluded that these aero-assisted maneuvers were more fuel efficient in most cases than conventional methods of changing orbital inclination.

A considerable amount of work in the area of pseudospectral optimization and its applications in orbital mechanics has been conducted by Dr Ross in his work at the Naval Postgraduate School. Ross and Hall [11] demonstrated an unusual approach to the orbit transfer problem involving the coupling of attitude dynamics and orbital mechanics in the development of a series of coplanar phasing maneuvers optimized for time, fuel, and control limitations. Their work incorporated the implementation of continuous thrusting into the optimal control problem. Dr Ross's work in this area culminated in the development of unique Zero-Propellant Maneuvers for the International Space Station [12]. These maneuvers utilize optimal control as well as feedback control techniques to take advantage of environmental conditions to minimize momentum saturation in the space station's control moment gyros. This development significantly decreased the cost of slewing the International Space Station.

2.3 Equations of Motion

The first step to solving any orbital mechanics problem involves developing a firm understanding in the dynamics inherent in the system. This involves first choosing a set of states to represent the system. The following sections detail three common choices of states in orbital mechanics and discuss their respective advantages and disadvantages.

2.3.1 The Two-Body Problem

The simplest problem in orbital mechanics is the Two-Body problem. This problem begins with two point masses and describes their mutual gravitational attraction to each other [4]. Vallado [13] mentions four fundamental assumptions made in the Two-Body problem:

1. The mass of the satellite is much smaller than the mass of the body it is orbiting. This allows the satellite's mass and its gravitational effects on the larger body to be neglected.

2. The frame of reference is inertial. This allows for derivatives to be taken without regard to the motion of the reference frame.

3. Both the Earth and the satellite are point masses.
4. No other forces are applied to either body.

These assumptions allow for the basic formulation of the Two-Body problem but constitute an imperfect model. One method for adjusting the model to account for these imperfections is known as Perturbation Theory. While the natural perturbations themselves are not discussed in this research, this theory can also be used to model maneuvers as perturbing accelerations.

The equations of motion are best described initially using an independent inertial coordinate frame as shown in Figure 2. In this figure, $\mathbf{R_C}$ denotes the position of the center of mass of the entire system. The vector $\mathbf{r}$ denotes the position of the second mass with respect to the system center of mass. The Equations of Motion for the second mass are:

$$\ddot{\mathbf{r}} = -\frac{G(m_1 + m_2)\mathbf{r}}{|\mathbf{r}|^3} \tag{1}$$

where G is the universal gravitational constant. Equation 1 can be simplified as follows:

$$\ddot{\mathbf{r}} = -\frac{\mu \mathbf{r}}{|\mathbf{r}|^3} \tag{2}$$

where $\mu = G(m_1 + m_2) \approx Gm_1$ is the specific gravitational constant for the system.

Since mass 2 is very small in comparison to mass 1, its gravitational effects on mass 1 can be neglected. This allows for the inertial frame to be moved to the center of mass 1 along with the center of mass of the system. In the case of a satellite orbiting the Earth, this yields what is commonly referred to as the Earth-Centered Inertial (ECI) reference frame. This reference frame consists of the three unit vectors [*i, j, k*]. The unit vector *i* is aligned with vernal equinox, *k* points to the North Pole, and *j* completes the right-handed system.

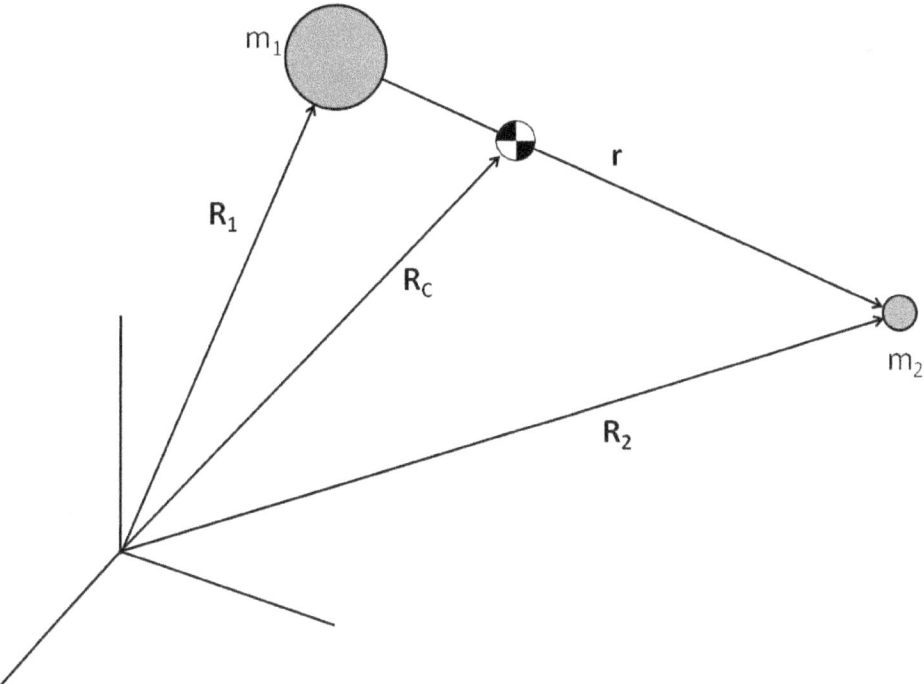

Figure 2: Two-Body Problem [4]

The Two-Body problem only accounts for the gravitational attraction between the two masses. There are various additional perturbing effects such as J2, which accounts for the oblateness of the Earth and air drag, that constantly influence the basic orbital motion of a satellite. Adding a perturbing acceleration, **A**, into Equation 2 yields the full equations of motion.

$$\ddot{\mathbf{r}} = -\frac{\mu \mathbf{r}}{|\mathbf{r}|^3} + \mathbf{A} \qquad (3)$$

If the acceleration being modeled is a maneuver, it is a function of only the thrust output and the mass of the satellite. Two cases are considered in this research: Continuous

Thrust and Impulsive Thrust. For the continuous-thrust case, the satellite mass can be modeled as constant due to the very low fuel consumption typical of electric propulsion, yielding a constant acceleration. For the Impulsive Thrust case, the fuel consumption is much higher and must be accounted for.

For a satellite undergoing constant acceleration, the resulting equations of motion in the ECI frame can be expressed as the following set of first-order derivatives.

$$\begin{bmatrix} \dot{x} \\ \dot{y} \\ \dot{z} \\ \dot{v}_x \\ \dot{v}_y \\ \dot{v}_z \end{bmatrix} = \begin{bmatrix} v_x \\ v_y \\ v_z \\ -\dfrac{\mu}{r^3}x + A_x \\ -\dfrac{\mu}{r^3}y + A_y \\ -\dfrac{\mu}{r^3}z + A_z \end{bmatrix} \qquad (4)$$

This method allows for a complete, closed form solution. However, due to their relative size, the Two-Body forces tend to dominate this formulation [14]. While numerical solvers today can handle the number of significant figures required to account for most perturbations, it is preferable to use a choice of states that change more slowly.

2.3.2 Classical Orbital Elements

Kepler's First Law states that orbital trajectories are conic sections with the attracting body at one of the foci. The Classical Orbital Elements (COE) represent a method of completely defining the orbit of a satellite with six parameters using conic section geometry. Figure 3 depicts the relationship between the six COE and the satellite's position and velocity. The COE are typically written as $(a, e, i, \Omega, \omega, v)$ where a is the semi-major axis, e is the eccentricity, i is the inclination, Ω is the right ascension of the ascending node (RAAN), ω is the argument of perigee, and v is the true anomaly. Depending on the application of the problem, the true anomaly may be replaced with the mean anomaly, M, or the eccentric anomaly, E. The following discussion on Classical Orbital Elements is taken from Wiesel [4, pp. 57-68].

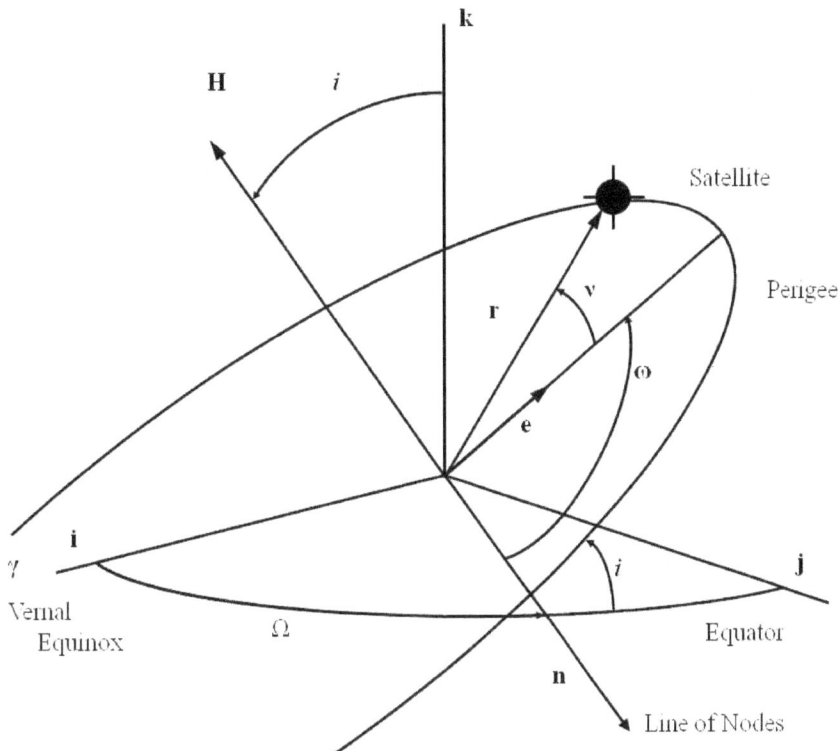

Figure 3: Classical Orbital Elements [4]

The semi-major axis, a, is defined as half the length of the longest axis on an ellipse and serves as a general measure of the size of an orbit as well as its orbital period. It is derived from the orbital energy of the satellite, ε.

$$\varepsilon = \frac{1}{2}v^2 - \frac{\mu}{r}$$
$$a = -\frac{\mu}{2\varepsilon}$$
(5)

Kepler's Second law states that the vector connecting the central body and the satellite will sweep out equal areas in equal times. This law led to the formulation of a quantity known as the mean motion, n defined as:

$$n = \frac{2\pi}{P} \qquad (6)$$

where P is the orbital period. Since the semi-major axis denotes the size of a stable orbit, it is related to the mean motion as shown in Equation 7. This Equation is known as Kepler's Third Law.

$$n = \sqrt{\frac{\mu}{a^3}} \qquad (7)$$

The eccentricity, e, is a measure of the orbital shape as shown in Table 1. For the purpose of this discussion, the circular and elliptical cases are all that will be covered. The eccentricity is determined from the magnitude of the eccentricity vector, **e**, which is calculated using the orbital angular momentum vector, **H**.

$$\mathbf{H} = \mathbf{r} \times \mathbf{v}$$
$$\mathbf{e} = \frac{1}{\mu}\left[(\mathbf{v} \times \mathbf{H}) - \frac{\mu \mathbf{r}}{r}\right] \qquad (8)$$

The inclination, i, measures the angle between the orbital plane and the inertial x-y plane. It is also calculated from the orbital angular momentum as shown in Equation 9.

$$\cos(i) = \frac{\mathbf{k} \cdot \mathbf{H}}{|\mathbf{H}|} \qquad (9)$$

Inclination is defined between 0° and 180°. Orbits in the 0° to 90° range are referred to as prograde orbits and are more commonly used than retrograde orbits, or those that occur between 90° and 180°.

Table 1: Eccentricity

Eccentricity	Shape
$e = 0$	Circular
$0 < e < 1$	Elliptical
$e = 1$	Parabolic
$e > 1$	Hyperbolic

The RAAN, Ω, measures the angle between the vernal equinox eastward to the line of nodes, **n**, shown in Figure 3 and calculated as follows:

$$\mathbf{n} = \frac{\mathbf{k} \times \mathbf{H}}{|\mathbf{k}||\mathbf{H}|} \tag{10}$$

The RAAN can be calculated by recognizing its relationship to the line of nodes.

$$\mathbf{n} = \cos(\Omega)\mathbf{i} + \sin(\Omega)\mathbf{j} \tag{11}$$

The argument of perigee, ω, denotes the location of the point on the orbit that is closest to the focal point at the center of the Earth. It is also calculated from the eccentricity vector and the line of nodes.

$$\cos(\omega) = \frac{\mathbf{n} \cdot \mathbf{e}}{|\mathbf{e}|} \tag{12}$$

For **e · k** > 0, ω can be directly obtained Equation 12 by taking the inverse cosine. However, if **e · k** < 0 then the inverse cosine function will yield an angle 180° from the true argument of perigee.

The first five COE denote the size, shape, and orientation of an orbit. The true anomaly, v, is a measure of where on that orbit the satellite currently resides. It can be calculated from the eccentricity and position vectors as shown in Equation 13.

$$\cos(v) = \frac{\mathbf{e} \cdot \mathbf{r}}{|\mathbf{e}||\mathbf{r}|} \tag{13}$$

Just as with the argument of perigee, **r · v** determines the quadrant for proper calculation of the true anomaly. The semi-major axis, eccentricity, and true anomaly may be directly related back to the magnitude of the position vector as shown in Equation 14.

$$r = \frac{a(1-e^2)}{1+e\cos(v)} \tag{14}$$

Despite the direct interpretation of the true anomaly, it is not always the best measure to use for orbital position [15]. The eccentric anomaly, E, is another measure of orbital position that is commonly used. The eccentric anomaly tracks the satellite's angular position on the orbit on a projected circle with equal radius to the semi-major axis as shown in Figure 4. This angle is measured from the center of the fictitious circle, O, rather than from the elliptical focal point, F. The eccentric anomaly is calculated from the eccentricity and the true anomaly.

$$\cos(E) = \frac{e+\cos(v)}{1+e\cos(v)} \tag{15}$$

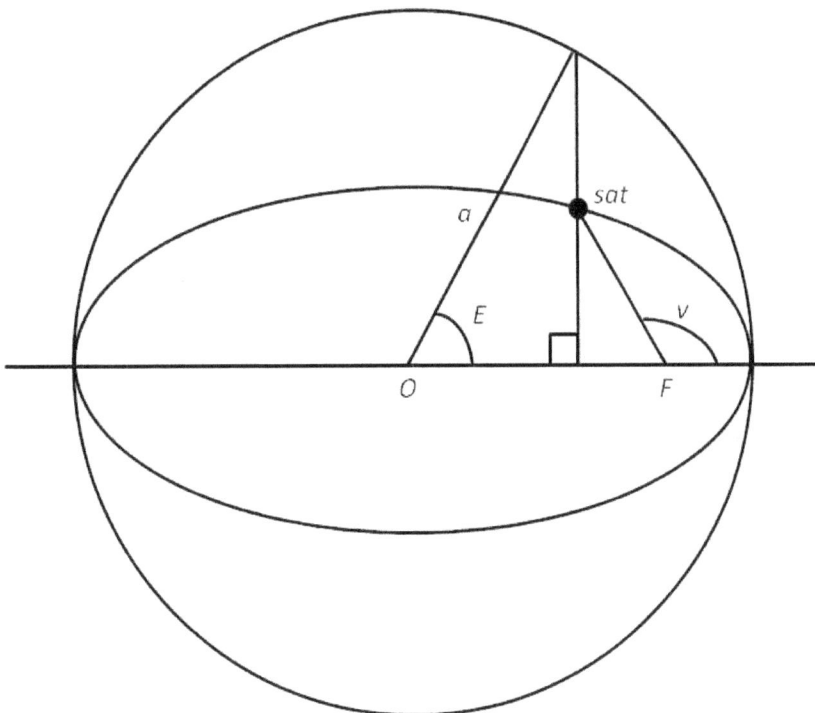

Figure 4: Eccentric Anomaly

The mean anomaly, M, is another common measure of orbital position created to assist in relating motion around an ellipse to motion around a circle. It relates directly to both the eccentric anomaly and the mean motion as shown in Equation 16.

$$M = E - e\sin(E) = n(t - t_0) \qquad (16)$$

In Equation 16, t_0 is the epoch time and t is time elapsed. It should be noted that at an eccentricity of zero, the mean, eccentric, and true anomalies are all equal.

For the basic Two-Body problem, five of the six COE are constant. When perturbations are added into the equations, these quantities change only due to the perturbing accelerations [14]. The Lagrange Planetary Equations (LPE) shown in Equation 17 govern how the COE change with these accelerations.

$$\begin{bmatrix} \frac{da}{dt} \\ \frac{de}{dt} \\ \frac{di}{dt} \\ \frac{d\Omega}{dt} \\ \frac{d\omega}{dt} \\ \frac{dM}{dt} \end{bmatrix} = \begin{bmatrix} \frac{2e\sin(v)}{n\sqrt{1-e^2}} & \frac{2a\sqrt{1-e^2}}{nr} & 0 \\ \frac{\sqrt{1-e^2}\sin(v)}{na} & \frac{\sqrt{1-e^2}}{na^2e}\left(\frac{a^2(1-e^2)}{r}-r\right) & 0 \\ 0 & 0 & \frac{r\cos(\omega+v)}{na^2\sqrt{1-e^2}} \\ 0 & 0 & \frac{r\sin(\omega+v)}{na^2\sqrt{1-e^2}\sin(i)} \\ -\frac{\sqrt{1-e^2}\cos(v)}{nae} & \frac{\sqrt{1-e^2}}{nae}\left(1+\frac{r}{a(1-e^2)}\right)\sin(v) & -\frac{r\cot(i)\sin(\omega+v)}{na^2\sqrt{1-e^2}} \\ \frac{(1-e^2)\cos(v)}{nae}-\frac{2r}{na^2} & -\frac{1-e^2}{nae}\left(1+\frac{r}{a(1-e^2)}\right)\sin(v) & 0 \end{bmatrix} \mathbf{A} + \begin{bmatrix} 0 \\ 0 \\ 0 \\ 0 \\ 0 \\ n \end{bmatrix} \quad (17)$$

where the perturbing acceleration $\mathbf{A}^\mathrm{T} = [A_r, A_t, A_n]$ denoting radial, tangential, and normal components, respectively [14].

Unlike the previous formulation, five of these six elements change very slowly. The sixth element, whether it be the true, eccentric, or mean anomaly, changes rapidly but in a predictable fashion. This method of defining an orbital state is intuitive but unfortunately has a number of singularities that tend to complicate the equations of motion. For instance, at zero inclination the RAAN loses meaning. Similarly, for zero eccentricity the argument of perigee becomes indistinguishable from the true anomaly. These singularities can be clearly seen in their equations of motion shown in Equation 17. Due to the location of these singularities, the COE are not necessarily the best set of states for numerical analysis.

2.3.3 Equinoctial Orbital Elements

A third method of completely defining an orbit is by the use of the Equinoctial Orbital Elements. This element set maintains the mathematical advantages of the COE without going singular for circular or prograde equatorial orbits. The following discussion on the Equinoctial Orbital Elements is taken from Kechichian [16].

This element set establishes another useful reference frame. The equinoctial reference frame is comprised of the unit vectors [*f*,*g*,*w*]. The unit vectors *f* and *g* span the orbital plane while *w* is aligned with the orbit angular momentum vector as shown in Figure 5.

The Equinoctial Orbital Elements may be derived directly from the COE. This change of variables is shown in Equation 18.

$$\begin{aligned}
a &= a \\
h &= e \sin(\omega + \Omega) \\
k &= e \cos(\omega + \Omega) \\
p &= \tan\left(\frac{i}{2}\right) \sin(\Omega) \\
q &= \tan\left(\frac{i}{2}\right) \cos(\Omega) \\
\lambda &= M + \omega + \Omega
\end{aligned} \tag{18}$$

The quantities h and k are the equinoctial reference frame components of the eccentricity vector. The quantities p and q relate the rotation from the ECI frame to the equinoctial reference frame as shown in Equation 19.

$$\begin{bmatrix} x \\ y \\ z \end{bmatrix}_{ECI} = \frac{1}{1+p^2+q^2} \begin{bmatrix} 1-p^2+q^2 & 2pq & 2p \\ 2pq & 1+p^2-q^2 & -2q \\ -2p & 2q & 1-p^2-q^2 \end{bmatrix} \begin{bmatrix} x_1 \\ y_1 \\ 0 \end{bmatrix} \tag{19}$$

Equinoctial Orbital Elements can be easily translated back into COE via the change of variables shown in Equation 20.

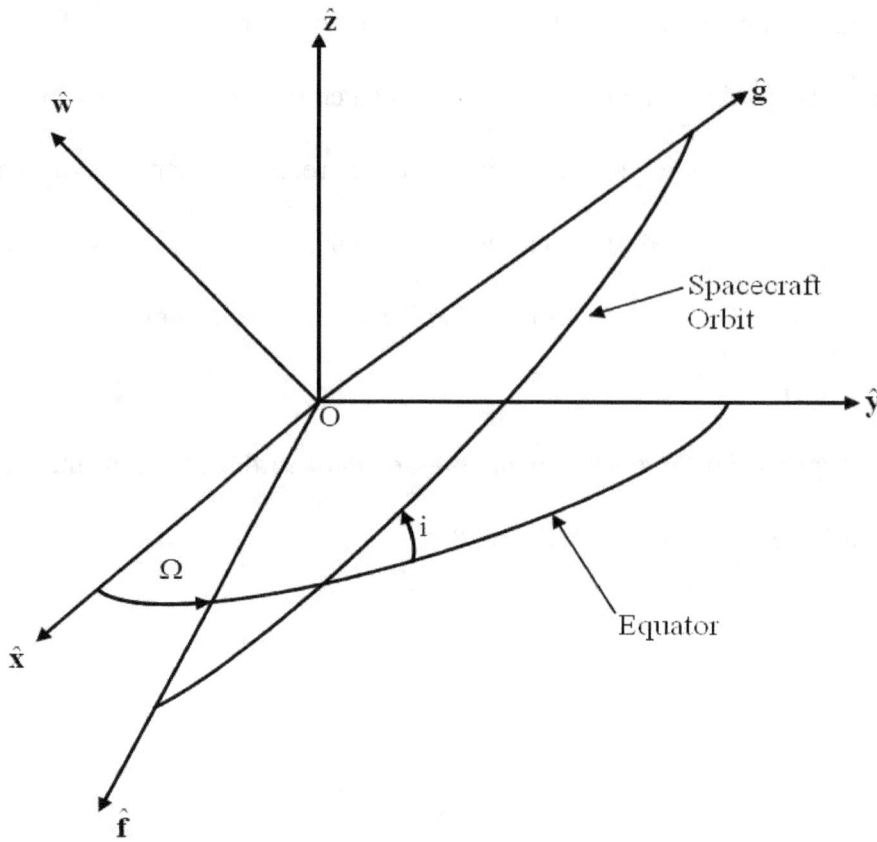

Figure 5: Equinoctial Reference Frame [16]

$$a = a$$
$$e = \sqrt{h^2 + k^2}$$
$$i = 2\tan^{-1}\sqrt{p^2 + q^2}$$
$$\Omega = \tan^{-1}\left(\frac{p}{q}\right)$$
$$\omega = \tan^{-1}\left(\frac{h}{k}\right) - \tan^{-1}\left(\frac{p}{q}\right)$$
$$M = \lambda - \tan^{-1}\left(\frac{h}{k}\right)$$

(20)

It can be seen from the conversion that while this new element set does not go singular for the circular or prograde equatorial cases, it does retain a singularity. Fortunately, this singularity occurs at an inclination of $180°$. Since retrograde equatorial orbits are rarely used, this singularity is of little concern in this research [13]. The position and velocity of the satellite in the Equinoctial Frame is given as:

$$\mathbf{r} = x_1\mathbf{f} + y_1\mathbf{g}$$
$$\dot{\mathbf{r}} = \dot{x}_1\mathbf{f} + \dot{y}_1\mathbf{g} \tag{21}$$

The components x_1, y_1, and their time derivatives from Equation 21 are defined as:

$$\begin{aligned} x_1 &= a\left[\left(1-h^2\beta\right)\cos(F) + hk\beta\sin(F) - k\right] \\ y_1 &= a\left[hk\beta\cos(F) + \left(1-k^2\beta\right)\sin(F) - h\right] \\ \dot{x}_1 &= a^2 nr^{-1}\left[hk\beta\cos(F) - \left(1-h^2\beta\right)\sin(F)\right] \\ \dot{y}_1 &= a^2 nr^{-1}\left[\left(1-k^2\beta\right)\cos(F) - hk\beta\sin(F)\right] \end{aligned} \tag{22}$$

where the quantities r and β are defined as:

$$\begin{aligned} r &= a\left(1 - k\cos(F) - h\sin(F)\right) \\ \beta &= \frac{1}{1+G} \\ G &= \sqrt{1-h^2-k^2} \end{aligned} \tag{23}$$

If the state vector is chosen as $\mathbf{z} = [a, h, k, p, q, \lambda]^T$ and the perturbing force is of the form $\mathbf{f} = f\mathbf{u}$ where $\mathbf{u}$ is a unit vector in the direction the force is being applied, then the state derivative follows the form:

$$\dot{\mathbf{z}} = \frac{\partial \mathbf{z}}{\partial \dot{\mathbf{r}}} \mathbf{A} + \mathbf{n} \tag{24}$$

Provided that the acceleration vector $\mathbf{A}$ is given in the equinoctial frame, the 3x6 matrix $\mathbf{M} = \frac{\partial \mathbf{z}}{\partial \dot{\mathbf{r}}}$ becomes:

$$\begin{bmatrix} \dfrac{2\dot{x}_1}{n^2 a} & \dfrac{2\dot{y}_1}{n^2 a} & 0 \\[6pt] \dfrac{G}{na^2}\left(\dfrac{\partial x_1}{\partial k} - \dfrac{h\beta \dot{x}_1}{n}\right) & \dfrac{G}{na^2}\left(\dfrac{\partial y_1}{\partial k} - \dfrac{h\beta \dot{y}_1}{n}\right) & \dfrac{k}{Gna^2}(px_1 - qy_1) \\[6pt] -\dfrac{G}{na^2}\left(\dfrac{\partial x_1}{\partial h} + \dfrac{k\beta \dot{x}_1}{n}\right) & -\dfrac{G}{na^2}\left(\dfrac{\partial y_1}{\partial h} + \dfrac{k\beta \dot{y}_1}{n}\right) & \dfrac{h}{Gna^2}(px_1 - qy_1) \\[6pt] 0 & 0 & \dfrac{Ky_1}{2Gna^2} \\[6pt] 0 & 0 & \dfrac{Kx_1}{2Gna^2} \\[6pt] \dfrac{1}{na^2}\left[-2x_1 + G\left(h\beta\dfrac{\partial x_1}{\partial h} + k\beta\dfrac{\partial x_1}{\partial k}\right)\right] & \dfrac{1}{na^2}\left[-2y_1 + G\left(h\beta\dfrac{\partial y_1}{\partial h} + k\beta\dfrac{\partial y_1}{\partial k}\right)\right] & \dfrac{1}{Gna^2}(qy_1 - px_1) \end{bmatrix} \quad (25)$$

In Equation 25, $K = 1 + p^2 + q^2$ and the partials of x_1 and y_1 with respect to h and k are given below in Equation 26.

$$\begin{aligned}
\dfrac{\partial x_1}{\partial h} &= a\left[-(h\cos(F) - k\sin(F))\left(\beta + \dfrac{h^2 \beta^3}{1-\beta}\right) - \dfrac{a\cos(F)}{r}(h\beta - \sin(F))\right] \\
\dfrac{\partial x_1}{\partial k} &= a\left[(h\cos(F) - k\sin(F))\dfrac{hk\beta^3}{1-\beta} + 1 + \dfrac{a\sin(F)}{r}(\sin(F) - h\beta)\right] \\
\dfrac{\partial y_1}{\partial h} &= a\left[(h\cos(F) - k\sin(F))\dfrac{hk\beta^3}{1-\beta} - 1 + \dfrac{a\cos(F)}{r}(k\beta - \cos(F))\right] \\
\dfrac{\partial y_1}{\partial k} &= a\left[(h\cos(F) - k\sin(F))\left(\beta + \dfrac{k^2 \beta^3}{1-\beta}\right) + \dfrac{a\sin(F)}{r}(\cos(F) - k\beta)\right]
\end{aligned} \quad (26)$$

Kechichian [17] stated that using F as the fast element rather than λ removes the requirement to solve Equation 16 at each integration step. This new set is known as a modified set of Equinoctial Orbital Elements. The equations of motion for F are given below in Equation 27.

$$\begin{aligned}
\dot{F} &= \dfrac{na}{r} + \dfrac{\partial F}{\partial \dot{\mathbf{r}}} f\mathbf{u} \\
\dfrac{\partial F}{\partial \dot{\mathbf{r}}} &= \dfrac{a}{r}\left[\dfrac{\partial h}{\partial \dot{\mathbf{r}}}\cos(F) + \dfrac{\partial k}{\partial \dot{\mathbf{r}}}\sin(F) + \dfrac{\partial \lambda}{\partial \dot{\mathbf{r}}}\right]
\end{aligned} \quad (27)$$

The partial derivatives of *h*, *k*, and λ are the second, third, and sixth rows of the matrix M as shown in Equation 25.

While the Equinoctial Orbital Elements avoid the singularities of the COE, the main disadvantage to using them is that from direct inspection it is not intuitively obvious what is happening physically to the system. The COE directly relate to the physical geometry of the orbit and as such are much simpler to directly interpret than the Equinoctial Orbital Elements.

2.4 Optimal Control Theory

The purpose of Optimal Control Theory is the determination of a time history of controls that satisfy the physical constraints of the system while minimizing or maximizing some performance criterion [18]. There are two primary categories of numerical methods for solving optimal control problems: Direct and Indirect Methods. Indirect Methods focus on derivation of first-order necessary conditions using the Calculus of Variations. These conditions are then used to pick a minimum cost extremal trajectory. Direct methods use Nonlinear Programming (NLP) techniques to satisfy a similar set of optimality conditions [19].

2.4.1 The General Indirect Method

The first step in Optimal Control Theory is establishing the problem. This consists of determining the equations of motion, cost function, and applicable constraints. The following brief explanation of terms is from Kirk [18]. The equations of motion can be written in first order form as:

$$\dot{\mathbf{x}}(t) = \mathbf{a}(\mathbf{x}(t), \mathbf{u}(t), t) \qquad (28)$$

where **x**(*t*) is an n-dimensional time history of the state vector and **u**(*t*) is an m-dimensional time history of the control vector. The symbols **x***(*t*) and **u***(*t*) below denote the optimal state and control vectors.

Constraints can be broken down into two primary types: path constraints and boundary constraints. Path constraints represent limitations on either the control or state at any time. For instance, engines have a finite amount of thrust yielding a maximum value for the control. It would be meaningless to solve for an optimal control solution that would require larger than the maximum available thrust. Boundary constraints pertain to either the final or initial states. They may be given as a set of equality or inequality constraints. A state vector that does not violate any constraint is referred to as an admissible trajectory. Similarly, a control vector that does not violate any constraint is referred to as an admissible control.

The cost function is generated by the designer and represents the quantities of importance.

$$J = h\left(\mathbf{x}(t_f), t_f\right) + \int_{t_0}^{t_f} g\left(\mathbf{x}(t), \mathbf{u}(t), t\right) dt \tag{29}$$

In the cost function, the function *h* is referred to as the Mayer term and denotes cost related to the final state. The function *g* is referred to as the Lagrange term or the running cost. This function tracks state and control costs that occur through their entire time histories. Cost functions may contain just the Mayer term, just the Lagrange term, or both depending on what is being optimized. Separate terms in the cost function are given appropriate weights designating their relative importance in the optimization. This is perhaps the most difficult part of designing the cost function. There are an infinite number of weighting combinations if multiple terms are present. As such, extreme care must be taken in properly balancing the relative weights in the cost function [18].

Equations 28 and 29 along with applicable constraints represent a complete optimal control problem. The first-order necessary conditions for optimality are derived using the Calculus of Variations:

$$\dot{\mathbf{x}}^*(t) = \frac{\partial \mathcal{H}}{\partial \lambda}$$

$$\dot{\lambda}^*(t) = -\frac{\partial \mathcal{H}}{\partial \mathbf{x}}$$

$$0 = \frac{\partial \mathcal{H}}{\partial \mathbf{u}} \tag{30}$$

$$0 = \left[\frac{\partial}{\partial \mathbf{x}} h\left(\mathbf{x}^*(t_f), t_f\right) - \lambda^*(t_f)\right]^T \delta \mathbf{x}_f$$

$$+ \left[\mathcal{H} + \frac{\partial}{\partial \mathbf{x}} h\left(\mathbf{x}^*(t_f), t_f\right)\right] \delta t_f$$

where $\mathcal{H}$ is the Hamiltonian constructed from Equations 28 and 29:

$$\mathcal{H}(\mathbf{x}(t), \mathbf{u}(t), \lambda(t), t) = g(\mathbf{x}(t), \mathbf{u}(t), t) + \lambda^T(t)\left[\mathbf{a}(\mathbf{x}(t), \mathbf{u}(t), t)\right] \tag{31}$$

In Equations 30 and 31, $\lambda(t)$ constitutes an n-dimensional vector of Lagrange multipliers, also known as co-states.

Boundary Conditions may be added to the problem formulation in Equation 30 as applicable. This research focuses on a fixed final time and free final state problem. Since $\delta \mathbf{x}_f$ is free, the fourth equation in Equation 30 results in:

$$\frac{\partial}{\partial \mathbf{x}} h\left(\mathbf{x}^*(t_f)\right) - \lambda^*(t_f) = \mathbf{0} \tag{32}$$

Conway [20] states that optimal control as applied to spacecraft trajectories run into several difficulties:

1. The dynamics are nonlinear.
2. Most practical trajectories are discontinuous.
3. The initial and final states may not be known explicitly.
4. Many of the forces such as planetary perturbations are time-dependent.
5. The basic structure of the trajectory may not be possible to specify *a priori*.

The use of low-thrust propulsion can alleviate the trajectory discontinuities since it can be used nearly continuously. This creates a very different problem from the traditional impulsive thrust model.

2.4.2 Primer Vector Theory

Primer vector theory is an indirect optimization method that satisfies the Necessary Conditions from Equation 30. Conway [20, pp. 16-20] describes the setup shown below for the problem of an optimal, constant specific impulse spacecraft trajectory. The conditions have been modified to conform to this research.

For a low-thrust engine, the acceleration can be constrained as $0 \leq A \leq A_{max}$. The cost function for the minimum fuel case with an additional Mayer term is:

$$J = h(\mathbf{x}(t_f)) + \int_{t_0}^{t_f} A\, dt \qquad (33)$$

In this case the ECI state vector is used:

$$\mathbf{x} = \begin{bmatrix} \mathbf{r} \\ \mathbf{v} \end{bmatrix} \qquad (34)$$

where **r** is the position vector and **v** is the velocity vector from the basic Two-Body problem discussed in Section 2.1.1. For this problem, the initial state $\mathbf{x}_0$ is specified. For this choice of states, the equations of motion are:

$$\dot{\mathbf{x}} = \begin{bmatrix} \mathbf{v} \\ \mathbf{g}(\mathbf{r}) + \Gamma \mathbf{u} \end{bmatrix} \quad (35)$$

where $\mathbf{g}(\mathbf{r})$ is the gravitational acceleration and **u** is the unit vector in the direction the thrust is being applied. The gravitational acceleration is modeled as shown in Equation 2. The Hamiltonian function can be constructed now as:

$$\mathcal{H} = \Gamma + \lambda_r^T \mathbf{v} + \lambda_v^T \left[\mathbf{g}(\mathbf{r}) + \Gamma \mathbf{u} \right] \quad (36)$$

The necessary conditions for the co-states are calculated from the Hamiltonian similar to the solution in Equation 30.

$$\begin{aligned} \dot{\lambda}_r^T &= -\frac{\partial \mathcal{H}}{\partial \mathbf{r}} = -\lambda_v^T \mathbf{G}(\mathbf{r}) \\ \dot{\lambda}_v^T &= -\frac{\partial \mathcal{H}}{\partial \mathbf{v}} = -\lambda_r^T \end{aligned} \quad (37)$$

In Equation 37, $\mathbf{G}(\mathbf{r})$ is the symmetric gravity gradient matrix given as:

$$\mathbf{G}(\mathbf{r}) = \frac{\partial \mathbf{g}(\mathbf{r})}{\partial \mathbf{r}} \quad (38)$$

The boundary condition is of similar form as Equation 32. This yields the following equations.

$$\begin{aligned} \lambda_r(t_f) &= \frac{\partial}{\partial \mathbf{r}(t_f)} h(\mathbf{x}(t_f)) \\ \lambda_v(t_f) &= \frac{\partial}{\partial \mathbf{v}(t_f)} h(\mathbf{x}(t_f)) \end{aligned} \quad (39)$$

The variables left are the acceleration magnitude, Γ, and direction, **u**. From inspection, it can be seen that the choice of direction that minimizes the Hamiltonian occurs when **u** is aligned opposite in direction to the velocity co-state, $\boldsymbol{\lambda}_v$. This term is referred to as the primer vector, **p**:

$$\mathbf{p}(t) = -\boldsymbol{\lambda}_v(t) \tag{40}$$

Conway [20] derives the primer vector equation from this definition.

$$\ddot{\mathbf{p}} = \mathbf{G}(\mathbf{r})\mathbf{p} \tag{41}$$

The boundary conditions for this differential equation come from Equation 39.

$$\begin{aligned}\mathbf{p}(t_f) &= -\frac{\partial}{\partial \mathbf{v}(t_f)} h(\mathbf{x}(t_f)) \\ \dot{\mathbf{p}}(t_f) &= \frac{\partial}{\partial \mathbf{r}(t_f)} h(\mathbf{x}(t_f))\end{aligned} \tag{42}$$

With this choice of **u** the Hamiltonian becomes:

$$\mathcal{H} = (1-p)\Gamma + \boldsymbol{\lambda}_r^T \mathbf{v} + \boldsymbol{\lambda}_v^T \mathbf{g} \tag{43}$$

From Equation 43 it can be seen that the Hamiltonian is a linear function of Γ. Therefore, the choice of acceleration magnitude is based on the sign of its coefficient. Conway [20] introduces the Switching Function to specify the acceleration magnitude.

$$S(t) = p - 1 \tag{44}$$

Here the choice of Γ comes from what Conway [20] refers to as the bang-bang control law:

$$\Gamma = \begin{cases} \Gamma_{max} & S > 0 \\ 0 & S < 0 \end{cases} \tag{45}$$

Note that this solution for Γ is singular if $S = 0$ for a finite length of time but will otherwise determine both thrust magnitude and direction for the specified optimal control problem. This analytical solution is very useful as a sanity check for the numerical solutions derived in later sections.

There are two primary advantages to using indirect methods: their relatively high accuracy and the absolute knowledge that they satisfy the first-order necessary conditions. However, they unfortunately have relatively small radii of convergence and require analytical derivations of the Hamiltonian. In addition, they also require a certain amount of *a priori* understanding of what the trajectory will look like. While direct methods are not as accurate as indirect methods, they do not suffer from the same disadvantages [19]. With the development and improvement of computer processing over the past half-century, these methods have become increasingly popular in solving optimal control problems without explicitly using the analytical necessary conditions [20].

2.4.3 Pseudospectral Methods

Pseudospectral techniques represent a class of direct methods that use collocation to solve optimal control problems numerically rather than analytically. This technique has become increasingly popular over the past several decades. The following discussion is taken from Conway [20, pp. 45-47] and Rao [21].

The first step is to discretize the state and control histories. This discretization is accomplished by the use of global polynomials. Discretization of the equations of motion is performed at collocation points. There are three sets of these points that are commonly used in pseudospectral methods: *Legendre-Gauss* (LG), *Legendre-Gauss-Radau* (LGR), and *Legendre-Gauss-Lobatto* (LGL). All three of these methods make use of the N-th order Legendre Polynomial, bounded on the interval [-1,1]:

$$P_N = \frac{1}{2^N N!} \frac{d^N}{dx^N}\left(\left[x^2 - 1\right]^N\right) \qquad (46)$$

The chief difference between these three methods is the inclusion or exclusion of the endpoints as shown in Figure 6. The LG points include neither set of endpoints, LGR points include only one set of endpoints, and LGL points include both sets of endpoints [22].

$$\begin{aligned} P_{LG} &= P_N \\ P_{LGR} &= P_N + P_{N-1} \\ P_{LGL} &= \frac{d}{dx} P_{N-1} \end{aligned} \qquad (47)$$

The boundary conditions for the differential equation for the LGL points are the endpoints. Note that there are two possible sets of LGR points, one set using the initial point and one using the terminal point. While similar in appearance, these three sets of points are distinctly different. Garg [23] proved that LG and LGR state and control solutions converge significantly faster than LGL and went on in [24] to demonstrate that LGR further improves accuracy. The pseudospectral method introduced in [24] was termed the Radau Pseudospectral Method (RPM) and is based on collocation using LGR points. The roots of the LGR polynomial form the set of discretization points for the RPM.

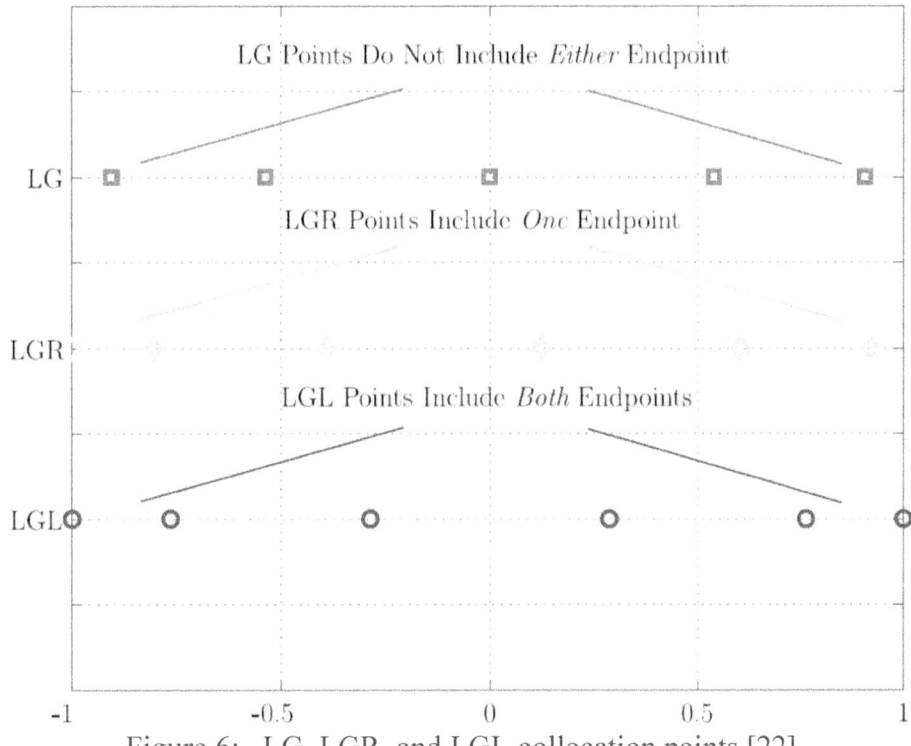

Figure 6: LG, LGR, and LGL collocation points [22]

It should be noted that there is a tradeoff inherent in this method of fitting points. A higher order polynomial will provide a better fit but will include more oscillations between each collocation point. While a lower order polynomial will fit the points less accurately, it will tend to be better conditioned, providing fewer oscillations between collocation points [20].

Once the states have been discretized and fitted with a polynomial, P(x) is differentiated. P'(x) is then compared to the defined state derivatives at the collocation points. The difference is referred to as the defect. These defects can be gathered into a vector as follows:

$$\Delta = [D]\mathbf{x}(t) - \mathbf{a}(\mathbf{x}(t),\mathbf{u}(t),t) \tag{48}$$

where [D] is the derivative matrix of the Legendre Polynomials. Figure 7 demonstrates this procedure for a single node. The defect then minimized in order to satisfy the specified equations of motion. Pseudospectral methods are generally known to converge spectrally. This means that convergence occurs faster than N^{-m} where N is the number of nodes and m is any finite value [21]. The numerical algorithm utilized in this research is based on the Radau Pseudospectral Method.

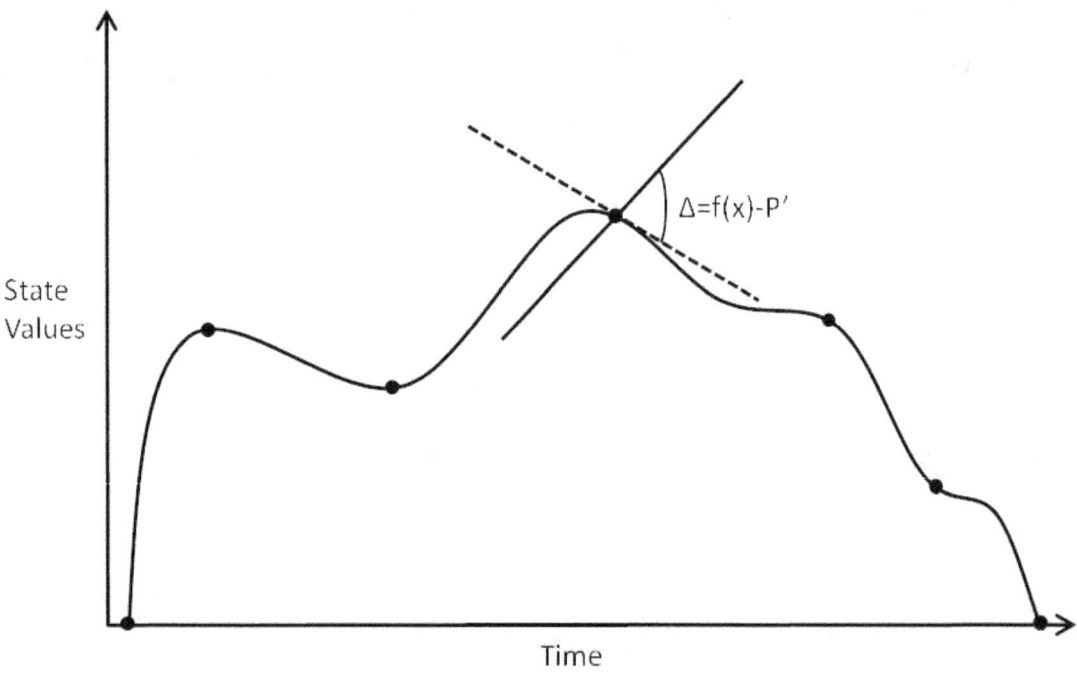

Figure 7: Defect Vector [20]

2.5 *Chapter Summary*

This chapter outlined the methodology behind the choices of states as well as the optimization methods used in this research. The Optimal Control Problem solved in the following chapters is conducted using equinoctial elements to avoid singularities but is translated back into classical orbital elements for analysis. The following chapter will outline in more detail the design and setup of the Optimal Control Problem.

III. Methodology

3.1 Chapter Overview

The following chapter outlines the methods used in this research in the development and execution of the Optimal Control Problem. The specific setup of the Optimal Control components is covered as well as an in-depth discussion of the software that was used in MATLAB®. Appendix A contains MATLAB® code that is discussed in this chapter.

3.2 Optimal Control Problem Formulation

This section describes the design and setup of the Optimal Control Problem. The equations of motion are specified along with their applicable state and control constraints. In addition, the design of the cost function is discussed in detail.

3.2.1 Equations of Motion

The modified Equinoctial Orbital Elements as discussed in Section 2.1.3 were selected as the states for this Optimal Control Problem. The corresponding equations of motion for this choice of states are outlined in Equations 24 through 27 in first-order form. The control variables were chosen as [T, θ, ψ] where T is the thrust magnitude, θ is the in-plane pitch angle shown in Figure 8, and ψ is the out-of-plane yaw angle. The resulting acceleration vector in the Equinoctial Reference Frame is given as:

$$\mathbf{A}^T = \frac{T}{m} \begin{bmatrix} (\sin(\theta)\cos(\phi) - \cos(\theta)\sin(\phi))\cos(\psi) \\ (\cos(\theta)\cos(\phi) + \sin(\theta)\sin(\phi))\cos(\psi) \\ \sin(\psi) \end{bmatrix} \qquad (49)$$

where *m* is the satellite mass and ϕ represents the satellite's position in the Equinoctial Reference Frame as shown in Figure 8. The angle ϕ is calculated from the components of the equinoctial position vector:

$$\phi = \tan^{-1}\left(\frac{y_1}{x_1}\right) \qquad (50)$$

where the quantities x_1 and y_1 are given in Equation 22.

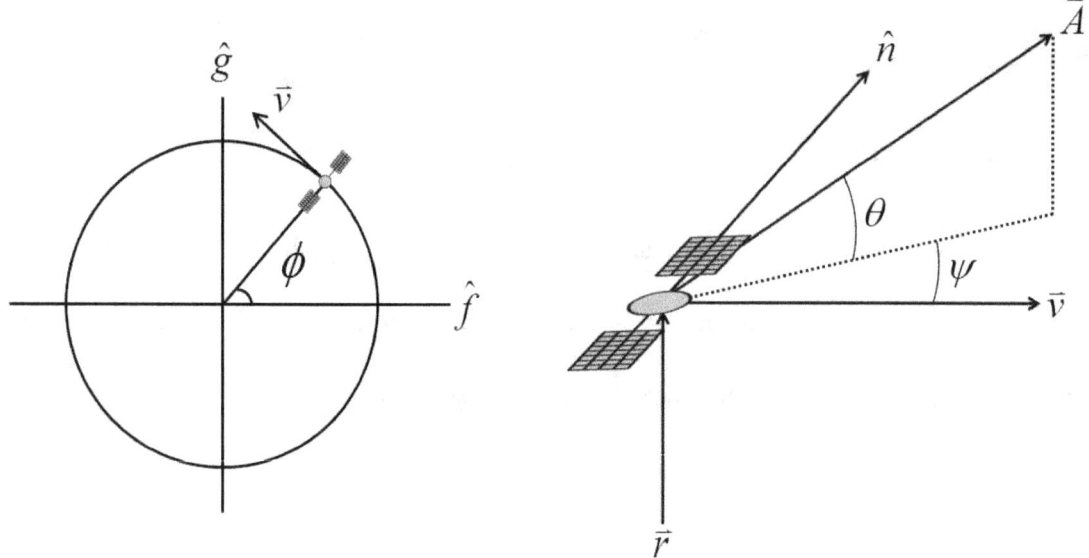

Figure 8: Thrust Vector

It should be noted that there are no perturbations to the basic Two-Body problem included in this realization of the equations of motion. Since the orbital trajectories of the reference and maneuvering satellites are nearly identical, it was assumed that the perturbation effects are also nearly identical. The position of the maneuvering satellite relative to its reference trajectory is one of the quantities of interest for this study and is incorporated into the cost function. Since the separation between the two trajectories is small, perturbation effects are not necessary to model and their absence allows for boosted efficiency in the numerical algorithms, decreasing run time significantly.

3.2.2 State and Control Constraints

Since most satellites generally wish to remain at the same altitude for mission requirements, the semi-major axis was constrained to a maximum deviation of 20 kilometers from the initial conditions. However, since the altitude component of the ellipse is only 10 kilometers and the intent is to maneuver as little as possible, this for all intents and purposes left the semi-major axis unconstrained.

The only constraint placed on eccentricity in this research was to assign it a maximum value of 0.5. This value was chosen in order to keep the code from incidentally generating a non-real value when using Equation 23 in the calculation of the state derivatives. The limits for the equinoctial elements h and k were determined from this restriction using Equation 18.

One of the goals of this research was to compare in-plane with out-of-plane maneuvers by leaving both as optimization parameters in this algorithm. As such, no restrictions were placed on inclination or RAAN. However, in order to bound the equinoctial elements p and q, it was assumed that their corresponding classical elements would only change by very small amounts using their relationship in Equation 18.

The final equinoctial element, F, is directly related to the mean anomaly and the argument of perigee. While the mean anomaly only increases over time, the argument of perigee changes rapidly at low eccentricity. The bounds applied to F were determined from extrapolating the final value of the mean anomaly of the reference satellite. Since the argument of perigee is bounded by $\pm\pi$ radians the bounds on F were established using its relationship to the mean anomaly and the argument of perigee. A summary of the global state constraints is given below in Table 2.

Table 2: Global State Constraints

State	Minimum	Maximum
a	$a(t_0)-20$	$a(t_0)+20$
h	-0.5	0.5
k	-0.5	0.5
p	$-2\tan(i(t_0)/2)$	$2\tan(i(t_0)/2)$
q	$-2\tan(i(t_0)/2)$	$2\tan(i(t_0)/2)$
F	$-\pi$	$\pi+F(t_f)$

The thrust magnitude was constrained in the code from zero to one denoting a full range from zero to full throttle. The MATLAB® function used for calculating the equations of motion was designed to scale this normalized throttle to a case-specific maximum thrust value.

The thrust angles were designed such that a single unique solution existed for virtually every thrust direction. The expected solutions for pitch angle were either velocity or anti-velocity and as such, specifying a limit from -180° to 180° was undesirable since it would result in a discontinuous solution for any optimal descending profile. Since pure altitude thrusting was assumed to be inefficient, the chosen singularity was placed at 270° for the pitch angle. The yaw angle was constrained from -90° to 90°. Since the satellite could thrust in any pitch direction, only half of a circle was required for the out-of-plane thrust angle. The applied control constraints are outlined in Table 3 below.

Table 3: Control Constraints

Control	Minimum	Maximum
T	0	1
θ	-90°	270°
ψ	-90°	90°

The initial iteration of the problem formulation applied a terminal event constraint to the Optimal Control Problem. This specified that the final position of the maneuvering satellite must be outside of the ellipse. This was accomplished using the formula:

$$E_{\text{In-Plane}} = \left(\frac{\Delta d}{a}\right)^2 + \left(\frac{\Delta h}{b}\right)^2 + \left(\frac{\Delta n}{c}\right)^2 \geq 1$$
$$E_{\text{Out-of-Plane}} = \left(\frac{\Delta n}{c}\right)^2 \geq 1 \tag{51}$$

where Δd is the in-track distance, Δh is the altitude difference, and Δn is the orbit normal distance between the reference and maneuvering satellites. The values a, b, and c denote the dimensions of the error ellipse in each of these directions. Due to the fuel inefficiencies inherent in out-of-plane maneuvering, a separate constraint was generated for this case in order to force the optimizer to converge on an out-of-plane maneuver. While this constraint ensured that the final positions would be outside of the ellipse, it tended to generate undesirable errors if the thrust magnitude or scenario time was insufficient for the maneuvering satellite to successfully exit the ellipse. Therefore, in subsequent versions of the code, the ellipse was applied as part of the cost function rather than as a constraint.

3.2.3 Cost Function

Two quantities were of interest in this research: fuel cost and the final position of the maneuvering satellite relative to its reference position. This necessitated both a Lagrange and a Mayer term in the cost function, written generically as:

$$J = B + \alpha \int_0^{t_f} T\,dt \qquad (52)$$

The Lagrange term, T, is the time history of the thrust magnitude, constituting the minimum thrust portion of the cost function. This term contains a weighting factor, α, that denotes the relative importance of minimizing fuel to ellipse avoidance. The primary purpose of the weighting factor was to balance the cost function such that the Mayer and Lagrange terms were on the same relative order of magnitude for each case. For the impulsive cases where the thrust time was small relative to the scenario time this required a weighting factor on the order of 1×10^{-2}. For the continuous case the thrust time was larger relative to the total scenario time requiring this weighting factor to decrease to the order of 1×10^{-6}. However, each case required specific manipulation of this variable in order to properly balance the cost function.

The Mayer term, B, is a three dimensional penalty function denoting an additional cost if the maneuvering satellite terminates inside the error ellipse. This method of representing the error ellipse was chosen in order to offset the undesirable results generated by the final state event constraint formulation of this problem. This penalty function would ideally be a Heaviside function, imposing the maximum penalty for any final state within the ellipse and no penalty for any final state outside of the ellipse. However, the derivative of a Heaviside function is discontinuous by definition and this problem required a function with a continuous derivative. Two smooth approximations were experimented with for the quantity B: an exponential form and a sigmoid penalty function as shown below.

$$B_{exponential} = \exp(-E)$$
$$B_{sigmoid} = \frac{1}{1+\exp(S \cdot (E-1))} \quad (53)$$

where E is the case-specific ellipse constraint as defined previously in Equation 51, and S is the desired sharpness of the sigmoid function. These functions were designed to approximate a Heaviside function, denoting large penalties when inside the ellipse and sharply dropping off as the maneuvering satellite departs the ellipse. The exponential form allows for increased control regarding how far outside the ellipse the designer wishes the satellite to travel. Figure 9 demonstrates the difference for a 2-D ellipse constraint between the two functions. Figure 9 (a) represents the relative weight imposed by an exponential function. Figure 9 (b) represents the relative weight generated by a sigmoid penalty function with S = 50. The weight in this figure is denoted by color with dark red representing the maximum penalty and dark blue representing the minimum penalty. The sigmoid penalty function was chosen for the results given in Chapter IV due to its decreased sensitivity to the weighting factor, α.

3.2.4 Multiple vs Single Phase Problem

The thrust profiles for the two impulsive cases were by their nature discontinuous. For this reason, an early attempt was made at separating thrusting and non-thrusting phases in the optimal control problem. This was accomplished by assigning three phases to the problem: two coasting phases and one thrusting phase. The problem was designed

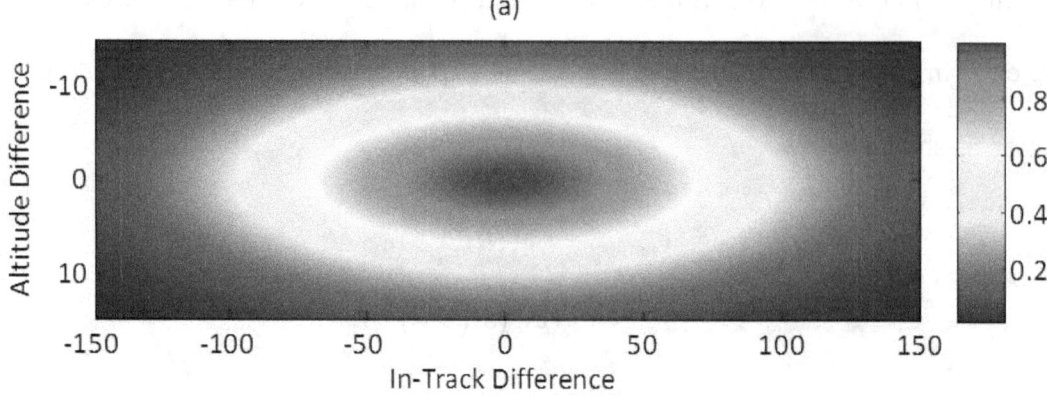

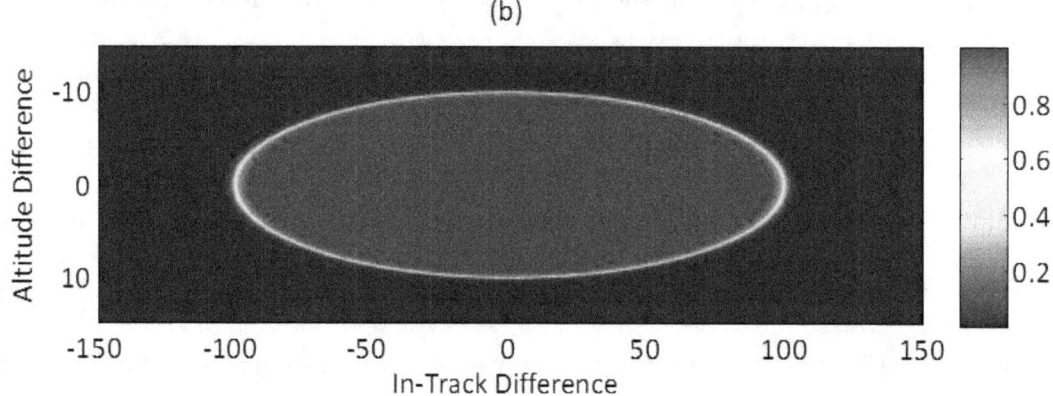

Figure 9: Penalty Functions

such that the thrusting phase was always the second phase with an associated event constraint that was used to force a minimum and maximum time to this phase in order to keep the solution within reasonable tolerances with respect to fuel expenditure. However, the existence of two phases independent of all three control variables yielded complications with convergence in GPOPS-II. For this reason, this attempt was abandoned and a formulation containing a single phase was designed that satisfied all three cases.

3.3 General Pseudospectral Optimal Control Software-II

GPOPS-II is an optimization software package created by Dr Anil Rao based on the Radau Pseudospectral Method as discussed in Section 2.4.3 and is specifically designed to run in MATLAB®. It incorporates an *hp*-adaptive mesh refinement algorithm for determination of the distribution of the collocation points [25]. It is designed specifically to work in conjunction with the nonlinear programming solvers IPOPT and SNOPT. The following is a summary of how GPOPS-II was utilized in this research to solve the Optimal Control Problem. For more detailed information on specific GPOPS-II functionality, see the GPOPS-II Manual [25].

3.3.1 Overview

The formulation of GPOPS-II involves a transformation from the standard method of describing the Optimal Control Problem discussed previously in Section 2.5.1 for a more generalized method. This method involves treating the Lagrange term in the cost function as a part of the Mayer term. This is valid once the Lagrange term has been integrated and is therefore only a function of initial and final time. When this transition is made, the integrand of the Lagrange term becomes another discretized vector in this formulation of the Optimal Control Problem. Any changes made to the state, co-state, and control history vectors during the optimization process generate an alternate integrand vector that subsequently changes the cost function.

In order to specify the Optimal Control Problem in GPOPS-II, several MATLAB® functions are required that define each component of the problem. These functions include but are not limited to:

1. Main code function
2. Continuous function
3. Endpoint function

In addition, upper, lower, and global limits must be specified for all variables manipulated within GPOPS-II. These limits are specified in MATLAB® through a complex array of structures [25].

3.3.2 Input Structure

Data is input to GPOPS-II through a single complex structure. Fields within this structure allow for everything from references to other required functions to an initial guess to limits on the states to be included in a single structure. The following is a summary of the input fields used in this research. The necessary substructures for the setup structure were summarized by Masternak [26] and are given in Appendix B.

The 'bounds' substructure specifies the upper, lower and global boundaries assigned to all variables manipulated within GPOPS-II. For the time limit field, minimum and maximum times at the scenario beginning and end may be specified, allowing for fixed or free initial or final time options within specified tolerances. Since this problem is fixed initial and final time, these minimum and maximum limits were identical. For the state and control limit fields, minimum and maximum bounds are placed on the initial, global, and final states in that order. This allows each state to be specified as either free or fixed at the endpoints as well as providing global restrictions to keep the state and control variables meaningful. In addition, each boundary condition or phase constraint that is expressed in a separate MATLAB® function requires a corresponding upper and lower bound to be specified in the limits substructure.

Like most optimization software, GPOPS-II requires an initial guess. This requires the user to have *a priori* knowledge of what the optimal solution should generally look like. Often, a poor guess can lead to convergence onto a suboptimal solution if the software determines the existence of a local minimum in the vicinity of the guess. Even without the presence of an additional local minimum to converge on, a poor initial guess can significantly increase the convergence time of the software.

Several additional MATLAB® functions must be specified for GPOPS-II to run properly. These functions are referenced under the 'functions' substructure as shown in Appendix B. Additional functions are optional depending on the problem statement but were not used in this research. The necessary components of these files are specified in later sections.

Not all subfields must be specified for proper functionality of GPOPS-II. One example is the optional 'mesh' subfield used in this research. This substructure allows for the user to specify settings for the *hp*-adaptive mesh. It may be used to place bounds on the number of desired collocation points as well as the criteria to set optimality and feasibility tolerances.

3.3.3 *Additional Required Functions*

The Continuous function is used to specify the quantities that are interior to the problem defined on an open interval (t_o, t_f) such as the equations of motion for the states. These quantities are read into the function via a complex input structure that contains the discretized vectors for each interior variable. This function may pass back three fields in its output structure: dynamics, path, and integrand.

The Endpoint function is used to specify conditions that apply to the boundaries of the Optimal Control Problem such as an event constraint. This function receives an input structure containing only the boundary values of each variable. Its output structure may contain two fields: objective and eventgroup. The objective field refers to the full Mayer term (including the integrated Lagrange term) of the cost function.

3.3.4 Output Structure

Upon convergence, GPOPS-II returns the calculated optimal solution through a single complex output structure. This structure includes but is not limited to state, co-state, control, and time histories. A complete list of the subfields to the GPOPS-II output structure is outlined in Appendix B.

3.3.5 Limitations

GPOPS-II has several important limitations inherent in its programming [21]. First, the states, controls, and co-states are assumed to be smooth. This was the reason for the choice of equinoctial elements as the states in this research as well as the requirement that the penalty function be continuous in the cost function. The lack of applicable discontinuities minimizes this problem. Second, despite the fact that the inequality path constraints are always satisfied at the collocation points, it is entirely possible for the constraints to be violated in between the collocation points. This problem is also minimized by the relatively loose constraints applied to this research and the use of an adaptive mesh inherent in the 'hp' method.

3.4 *Systems Tool Kit® v 10*

Systems Tool Kit® (STK) v 10 is a software geometry engine designed by Analytical Graphics Incorporated® (AGI) in order to display dynamic positions and attitudes of space vehicles. It was utilized in this research both as the engine to generate realistic scenario data as well as the method of visualizing and verifying the optimal thrust solution. Access between MATLAB® and STK was accomplished through the built-in Component Object Model (COM) Interface. This tool allowed for direct control of virtually all STK functionality from within MATLAB® using a complex structure of handles. Appendix A contains a library of reference functions that were designed specifically for use in this research in order to better facilitate communication between these two programs.

3.4.1 *Component Object Model Interface Library*

A library of functions was designed as part of this research in order to facilitate direct control of STK from within MATLAB®. This library utilized the COM interface in order to establish an active communication pathway to MATLAB®. This interface was created specifically for the purpose of providing users with the ability to control and automate objects within STK and requires the STK/Integration Module license in order to operate [27].

The COM interface facilitates external control for compatible programs using a series of handles. These handles are structures containing pointer variables that access specific objects in the active program. The most important handles used in this code are the User Interface Application (uiapp) and Object Model Root handles. The uiapp handle serves as the variable that tracks the COM Automation server default interface between the two programs. If at any time this variable is deleted or overwritten, the program is closed and all related handles are released. The root handle can be obtained from the uiapp handle via its 'Personality2' subfield. Objects within STK can then be directly manipulated through the COM interface using subfields contained within the root handle [27].

Each function in this library was designed to complete a specific task in STK and relay the relevant Object Model handles back to MATLAB®. These functions are all designed generically with no scenario-specific information included. This was conducted such that the scenario-specific data could be housed in the main MATLAB® code, allowing this library to be useful for future research in this area. Table 4 below details the name and purpose of each of the functions in the STK COM Interface Library.

3.4.2 Scenario Input

The main code for this research began by initializing STK and designating the scenario start time. The chosen scenario was set to occur on 1 Jan 2013 at 0900. Once the scenario was created the code automatically generated the appropriate area target. The parameters for the area target are given below in Table 5. This location is also shown below in the STK 2D plot in Figure 10.

Both a reference and maneuvering satellite were then created in STK with identical initial conditions. The COE sets shown in Table 6 were used for these initial conditions. These two sets of initial conditions were chosen in order to explore the differences between single orbit reentry into the AOR versus a multiple orbit scenario.

The satellite dry mass used was 400 kg with 100 kg of on-board fuel for a total satellite wet mass of 500 kg. A 0.5 N Electric Propulsion thruster was used for the continuous thrust case while the impulsive thrust case utilized a 22 N thruster.

Table 4: STK COM Interface Library Function List

Function	Description
Area_Target	Creates an area target object
Astrogator	Creates a satellite object in utilizing the Astrogator engine to propagate maneuvers
Compute Access	Generates an access report between two objects
Create_Engine_Model	Creates a custom engine model in the Component Library
Elements	Calculates the orbital element time history for the specified satellite object
FTV_Maneuver	Generates a Finite Thrust Vectored maneuver in the Maneuver Control Sequence (MCS) in Astrogator
Initialize	Opens new STK window and automatically fills general scenario information
ITV_Maneuver	Generates an Impulsive Thrust Vectored maneuver in the Maneuver Control Sequence (MCS) in Astrogator
Maneuver_From_File	Generates a Finite Thrust Vectored maneuver in the Maneuver Control Sequence (MCS) in Astrogator utilizing an external text file for attitude control
Output_to_text	Generates a text file conforming to the *.a thrust attitude external file input parameters
Propagate	Adds propagation step in the Maneuver Control Sequence in Astrogator

Table 5: Area Target Parameters

Location	Latitude	Longitude	Radius	Min elevation angle
AFIT	39.783 N	275.917 W	500 km	20°

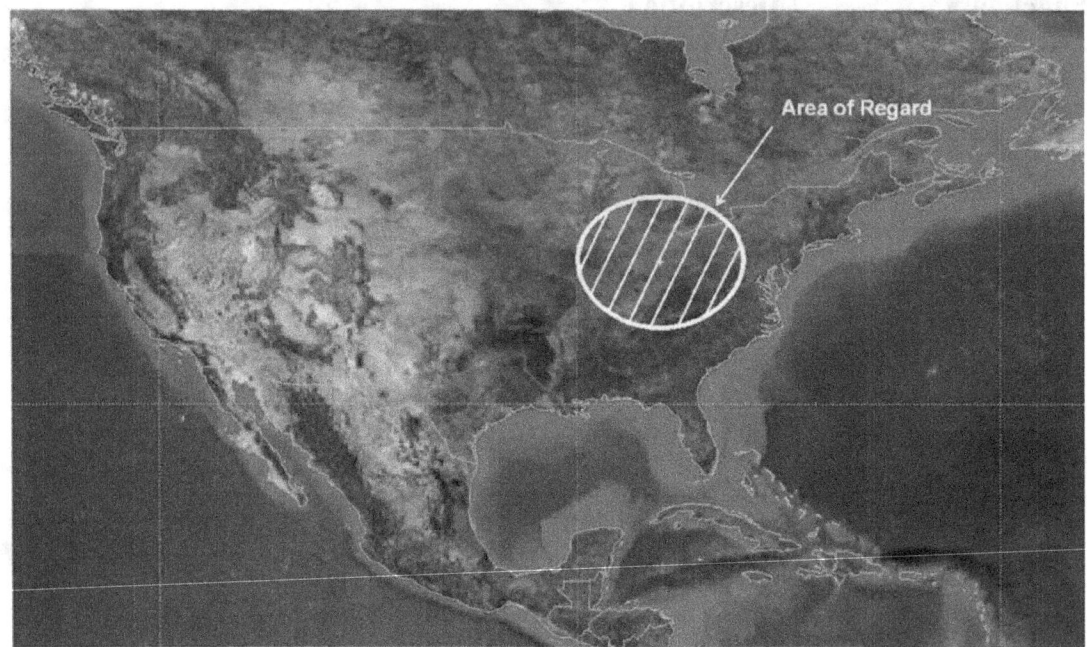

Figure 10: Area of Regard

An error ellipse was then generated around the reference satellite with a semi-major axis of 100 kilometers oriented along the velocity vector and semi-minor axes of length 10 kilometers denoting altitude and distance along the reference satellite's orbit normal vectors. This error ellipse is shown in the STK 3D plot in Figure 11.

Using the given initial conditions, STK then generated an access report between the area target and reference satellite for each scenario. This report was imported into MATLAB® to determine the first AOR departure time. This time served to account for the coast time from the specified STK scenario epoch until the optimzation start time.

The access report also generated the subsequent AOR reentry time which served as the optimization scenario termination time. With this information, GPOPS-II was able to solve the optimal control problem.

Table 6: Satellite Initial States

	Altitude	Eccentricity	Inclination	RAAN	Argument of perigee	True anomaly
1	500 km	10^{-6}	45°	0°	0°	60°
2	500 km	10^{-6}	45°	50°	0°	60°

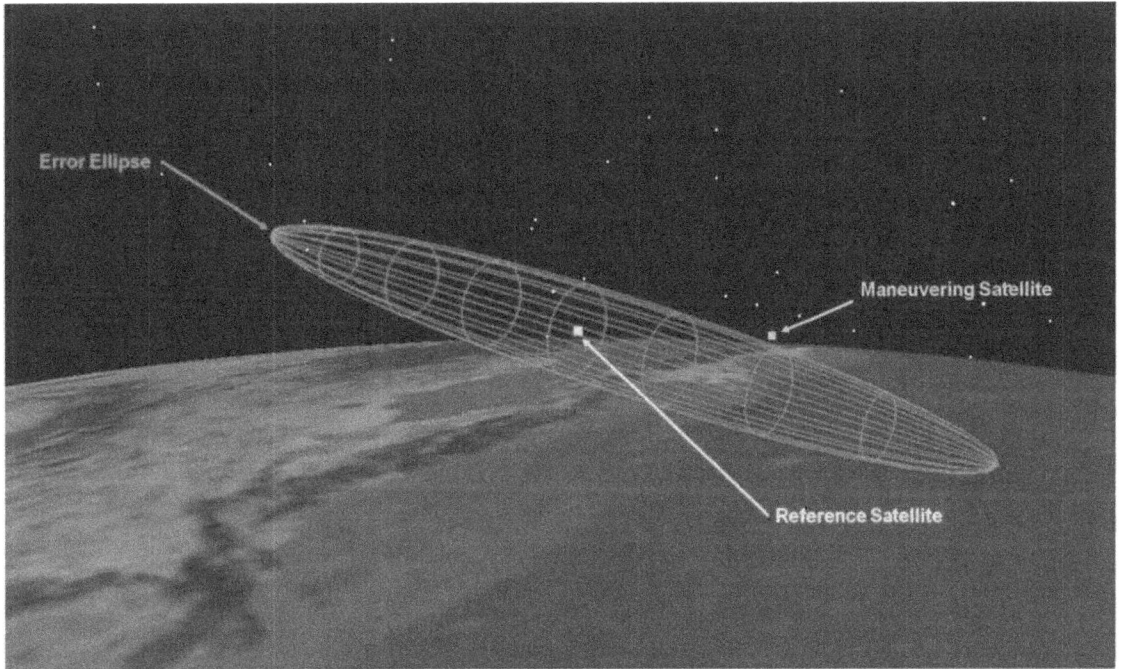

Figure 11: Error Ellipse

3.4.3 Maneuver Development

Upon convergence in GPOPS-II, the commanded thrust profile was uploaded into STK. The maneuvering satellite was then commanded to execute the calculated profile. Orbital element reports were generated via STK for the resulting trajectories based on the

COE history of both the reference and maneuvering satellites. The flowchart shown in Figure 12 depicts the interface between MATLAB® and STK. The dotted line on this figure demonstrates a critical step in the design process for this algorithm. Inconsistencies in the output from GPOPS-II and STK were compared and additional test runs were conducted using alternate GPOPS-II settings in order to refine the solution.

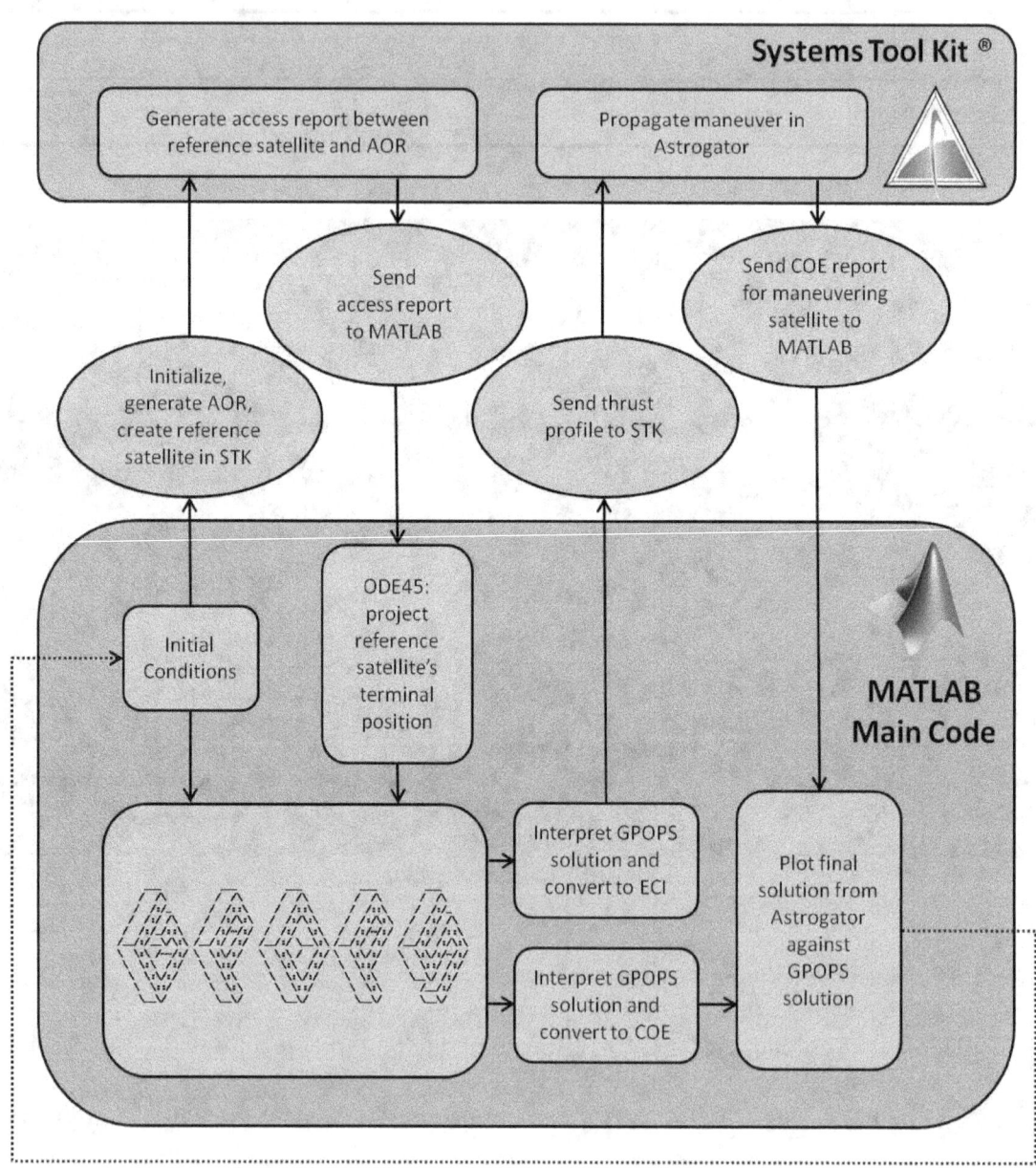

Figure 12: Optimization Routine Flowchart

One of the most powerful capabilities that STK added to this code was the ability to easily simulate the effects of perturbations to ensure that they did not interfere with ellipse avoidance generated by the thrust profiles being calculated. As was previously mentioned, the code within GPOPS-II only ran the Two-Body equations of motion since the reference and maneuvering satellites were in such close proximity for the entire scenario. However, once the profile had been uploaded to STK, perturbations could be easily added back into the scenario in order to visualize their effects on the calculated trajectories.

3.4.4 Optimizer Result Validation

In addition to data collection and visualization, STK was utilized in this research in order to validate the maneuver results from GPOPS-II. This was accomplished using a basic parameter search on the control variables. This search was conducted in MATLAB® and utilized the STK COM Interface Library in order to input a large variety of potential maneuvers and compare their relative cost as defined in Section 3.2.3.

For the impulsive parameter searches, pitch and yaw were varied in accordance with the constraints given in Table 3. In each of these cases, the pitch angles were varied while thrust magnitude and duration were held constant. Since this yielded identical delta v costs for each of these maneuvers, the particular value of interest in the cost function then becomes ellipse avoidance. The ellipse constraint from Equation 51 was then utilized to evaluate the relative value of each combination of pitch and yaw angles. These relative values were visualized using the *imagesc* command in MATLAB®. This command visually illustrates the relative sizes of elements in a matrix using color coding.

For the thrust duration parameter search, the thrust time was varied from 80% to 120% of the GPOPS-II solution. At each value of thrust duration, ellipse avoidance was calculated in identical fashion to the pitch and yaw angle parameter search.

For the continuous single orbit parameter search, the pitch angle was varied by both translating ±10° and skewing 80% to 120% from the GPOPS-II solution. After the pitch angle solution was perturbed, the same ellipse avoidance calculation was conducted as with the thrust angle and duration parameter searches. The *imagesc* command was also used in this case to visualize the result of perturbing the pitch angle solution. Due to the relatively short thrust duration for the continuous multiple orbit scenario, it was treated as an impulsive case for this analysis.

3.5 Chapter Summary

This chapter outlined the setup of the Optimal Control Problem, the design of the problem within GPOPS-II, and the implementation of STK in determining and validating the solution. The next chapter will discuss the results returned by GPOPS-II and the analysis of those results when executed in STK.

IV. Analysis and Results

4.1 Chapter Overview

This chapter outlines the results of the three test cases described in Chapter I using the algorithm developed in Chapter III. The first portion of this chapter describes the solution to the optimal control problem for each of the three cases as determined by GPOPS-II. The next portion of this chapter outlines the results from Systems Tool Kit® when these maneuvers are input from the GPOPS-II code for validation and proof of concept.

4.2 Optimal Control Results

This section presents the optimal thrust results for each of the three cases outlined in Chapter I: Impulsive In-Pane, Impulsive Out-of-Plane, and Continuous In-Plane. For the first and third cases, two families of solutions exist. The first solution is to the scenario in which the satellite only takes one orbit from AOR departure until re-entry. The second solution consists of multiple orbits between AOR departure and subsequent re-entry. The size and geographic location of the specified AOR will dictate how frequently this second scenario occurs. However, even for a relatively small AOR the single orbit scenario is easily the most common. The results presented in this section represent the Two-Body approximations calculated in GPOPS-II. The maneuvers from this section are tested in STK with perturbations in Section 4.3.

Each solution presented for Cases 1 and 3 represent a desire to climb when maneuvering. There is a corresponding solution that allows for a descent in both of these cases. Case 2 also contains two possible solutions depicting thrusting in either orbit normal direction. These additional solutions have been excluded from this section due to redundancy. For each case, a short coast time occurs at the beginning of each solution. This is the result of beginning the STK scenario prior to AOR entry.

For all three cases, a convergence tolerance of 1×10^{-8} was set for the adaptive mesh in GPOPS-II. The optimizer was allowed a maximum of 45 mesh iterations in order to converge to this tolerance. Each case required manipulation of the weighting factor as previously discussed in Section 3.1.3 as well as manipulation of the initial number, distribution, and iterative increment of collocation points. The nodal distribution required adjustment in each case due to scenario length and complexity. The default nodal distribution in GPOPS-II is ten segments with four nodes per segment. However, due to the length of time between each node, an increase in the number of total points in the state history in GPOPS-II was required. For this reason, the single orbit nodal distributions are smaller than the multiple orbit nodal distributions.

4.2.1 Case 1 Single Orbit

For the Impulsive In-Plane single orbit scenario, the satellite was given the first set of initial conditions specified in Table 6 in Chapter III. For this case the weighting factor was set at 9×10^{-3}. The optimizer started with ten segments containing seven nodes per segment and was allowed to increase the nodes in each segment at a range from 20 to 25 points per mesh iteration. The optimized thrust profile for this scenario is shown in Figure 13 with the resulting Two-Body orbital elements for the maneuvering satellite given in Figure 14.

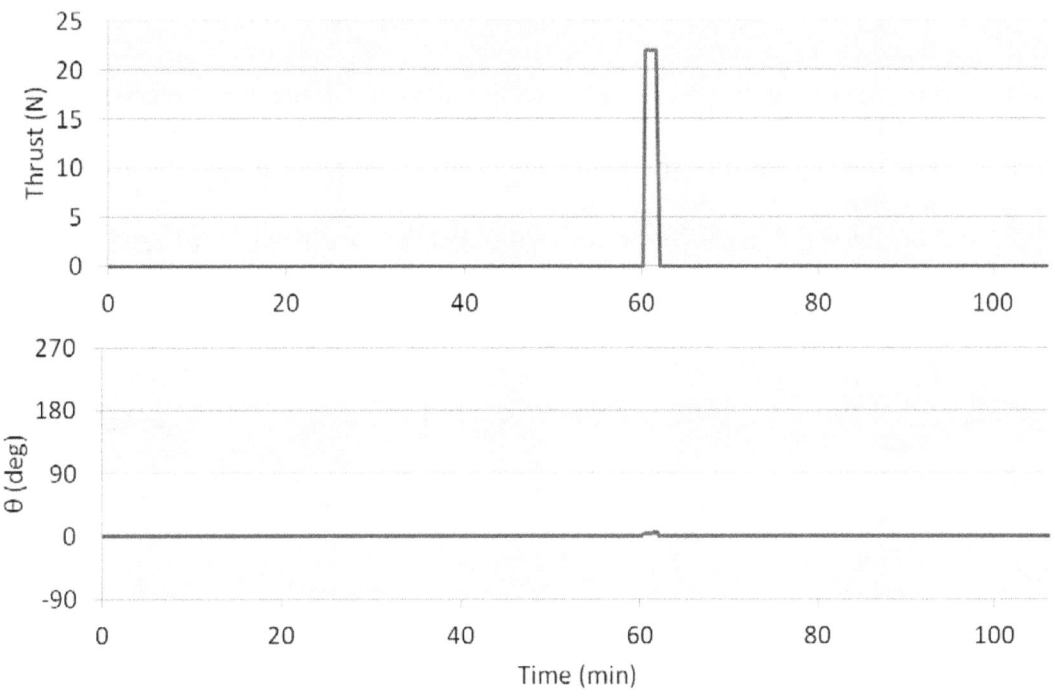

Figure 13: Case 1 Single Orbit Thrusting Profile

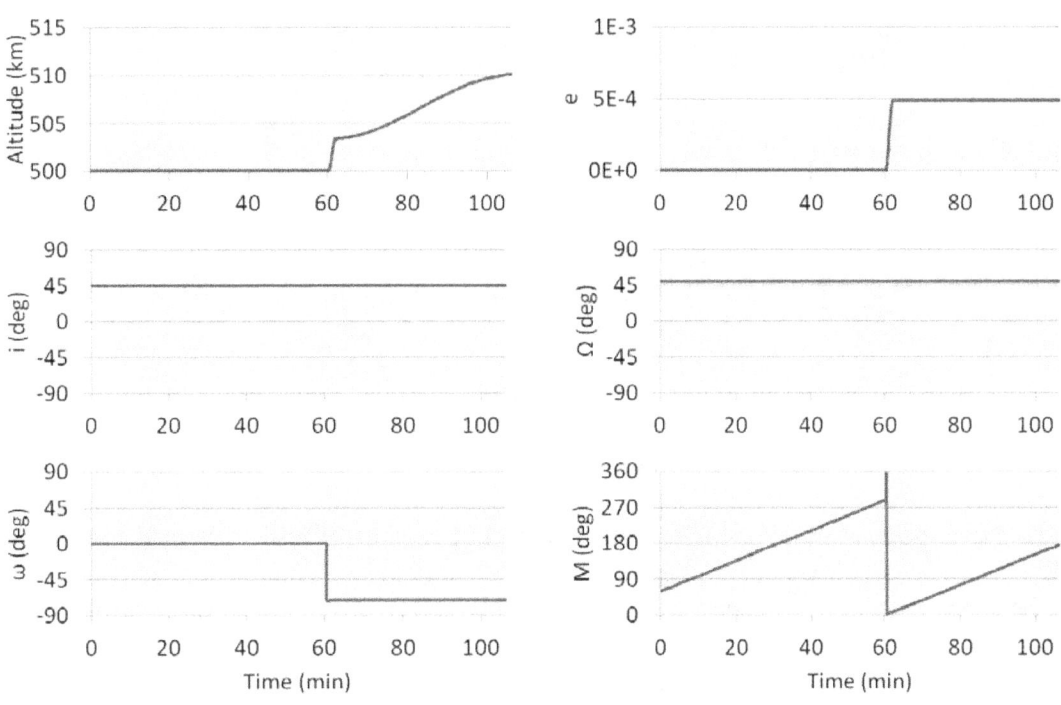

Figure 14: Case 1 Single Orbit COE

To better understand the relative motion between the two trajectories, the orbital elements were converted to give a cross-sectional view of the error ellipse shown in Figure 15. In this figure, the solid line represents the trajectory of the maneuvering satellite and the dashed line shown in this figure depicts the ellipse. This reference frame is fixed with the current position of the non-maneuvering trajectory always at the origin.

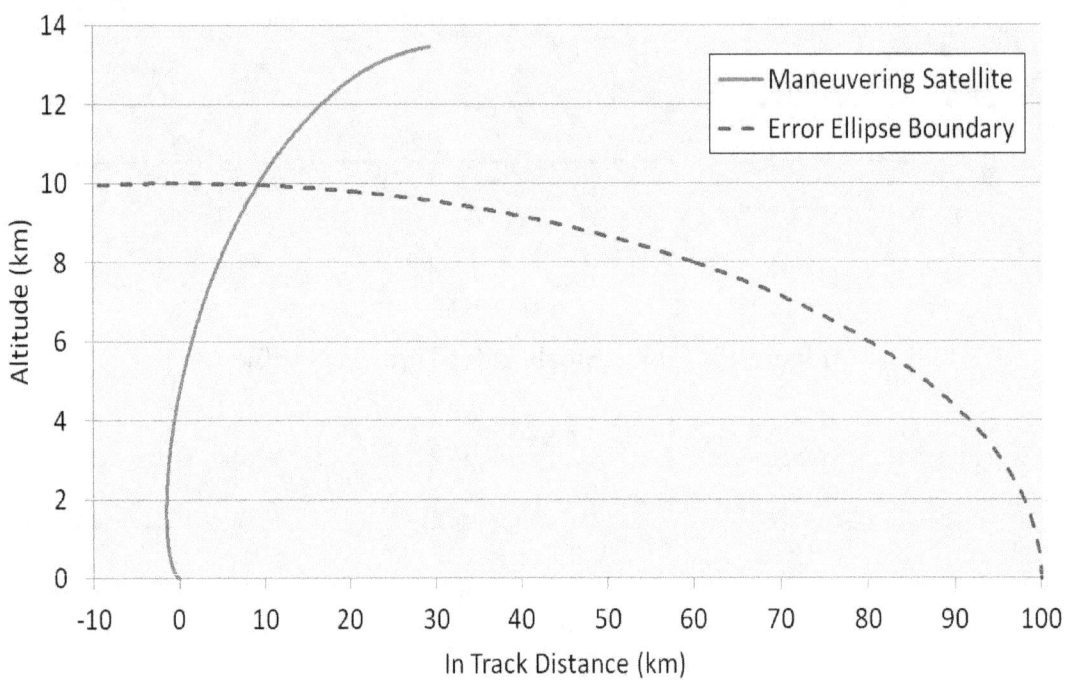

Figure 15: Case 1 Single Orbit Relative Motion Cross Section

Due to the short duration of the scenario, insufficient time is available for the satellite to achieve significant in-track spacing. Therefore, this solution represents intent to use a change in altitude as the primary method to exit the ellipse. From this solution, it can be seen that the best place to insert an impulsive thrust is half an orbit prior to AOR reentry. This maneuver effectively places apogee at the final position as shown in Figure 15 and uses the change in the semi-major axis to maneuver out of the ellipse. The calculated fuel cost for this maneuver is approximately 3.7 m/s.

4.2.2 Case 1 Multiple Orbit

The second set of initial conditions from Table 6 defines the multiple orbit scenario for Case 1 as discussed in this section. For this scenario, the weighting factor was set at 1 x 10^{-2}. The multiple orbit scenario consisted of a much smaller impulse relative to the overall scenario time, requiring the initial number of collocation points to be initially increased to 25 nodes per segment in order to obtain a solution with finer resolution. The number of nodes added per mesh iteration was also increased to a range of 20 to 25. The optimized profile for this scenario is shown below in Figure 16. The resulting Two-Body orbital elements for this solution are given in Figure 17 and its cross sectional plot is given in Figure 18.

This scenario demonstrates that if multiple orbits are expected to occur prior to AOR reentry it is advantageous to thrust early. Even a small initial change in semi-major axis creates an difference in orbital period that when propagated over the approximately 17 hour scenario will allow for a large enough in-track spacing between the maneuvering satellite and its projected reference trajectory to escape the ellipse. This maneuver can therefore be accomplished with a much smaller impulse than the single orbit scenario. The altitude change completed in this scenario is approximately 2 km rather than the nearly 14 km of altitude change observed from the single orbit scenario. Figure 18 demonstrates the path this maneuver takes to exit the ellipse. As this figure demonstrates, the slightly larger orbital period allows for long-term divergence between the two trajectories to increase the in-track spacing. This maneuver has an approximate delta v requirement of 0.6 m/s.

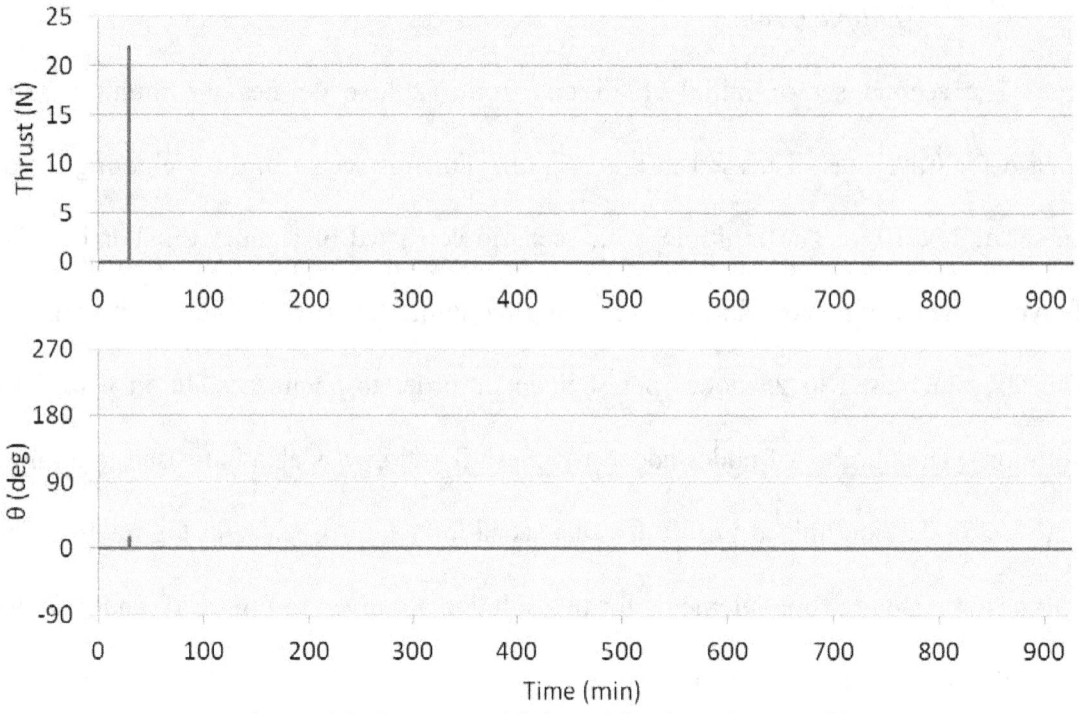

Figure 16: Case 1 Multiple Orbit Thrusting Profile

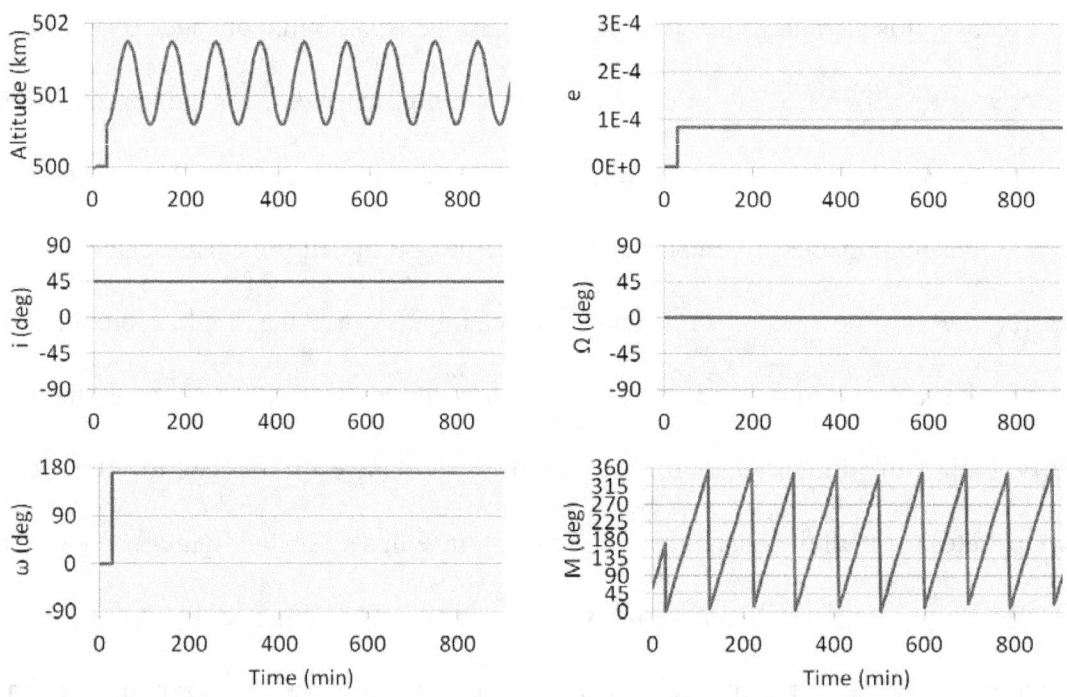

Figure 17: Case 1 Multiple Orbit COE

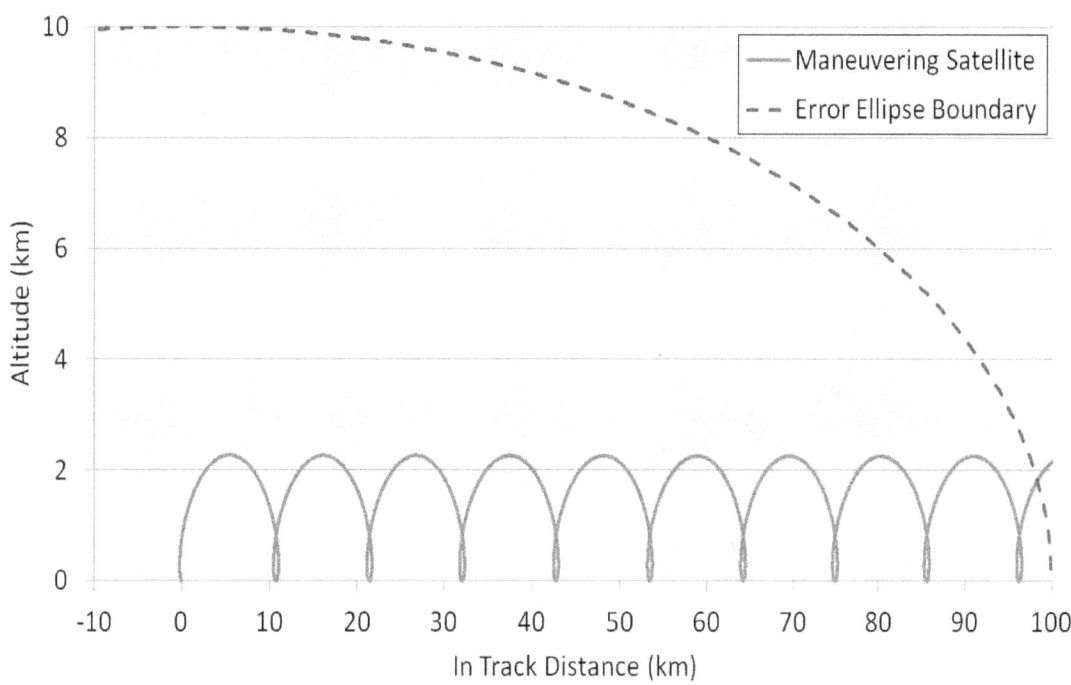

Figure 18: Case 1 Multiple Orbit Relative Motion Cross Section

4.2.3 Case 2 Single Orbit

As discussed in Section 3.2, the cost function was modified for the Impulsive Out-of-Plane case according to Equation 51 in order to remove any advantage to maneuvering for either altitude or in-track spacing from the terminal cost. This modification to the cost function allowed the software to converge on an optimal out-of-plane maneuver. Since this solution required modification of the Mayer term in the cost function in order to converge, the Case 2 profile is by no means globally optimal. However, this solution provides other advantages that are discussed later in this chapter.

The satellite was given the first set of initial conditions shown in Table 6. The weighting factor was set at 1×10^{-4} for this scenario. The number of collocation points was initially set at 4 nodes per segments and was increased between 4 and 10 nodes per mesh iteration. The optimal thrust profile for this case is shown in Figure 19. Due to the

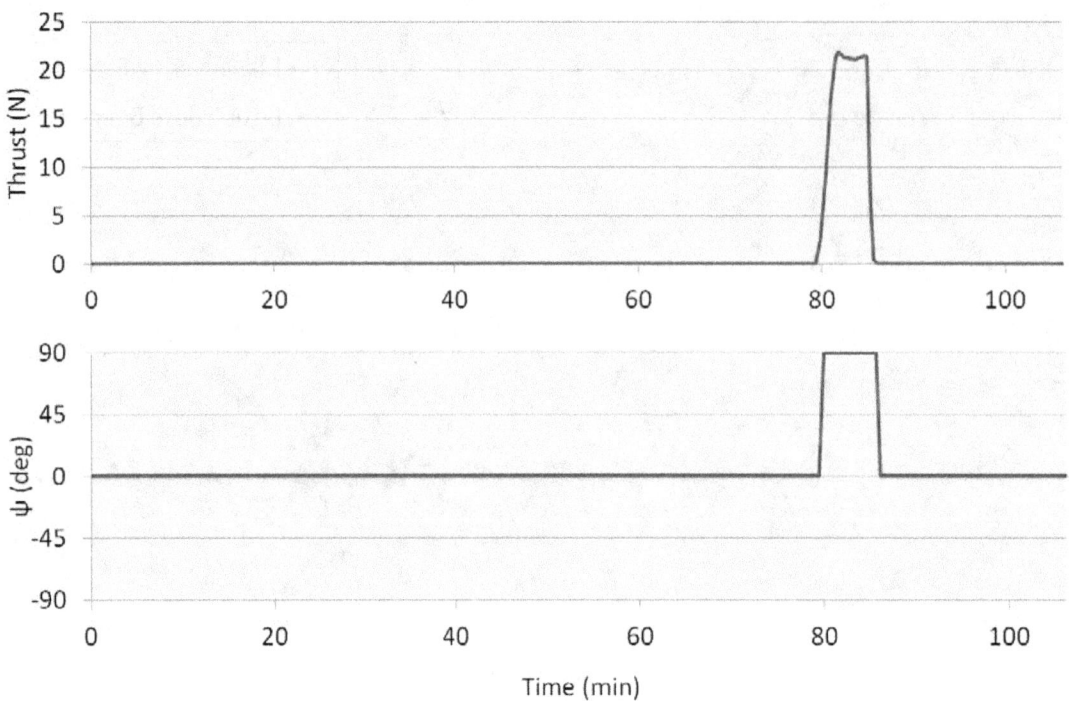

Figure 19: Case 2 Thrusting Profile

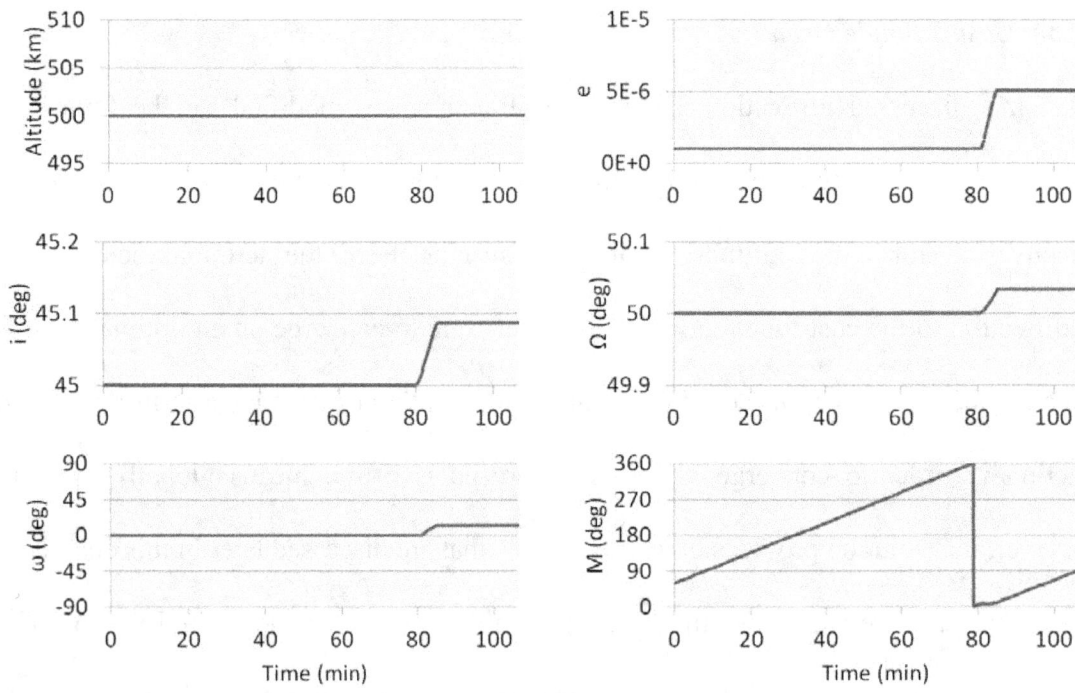

Figure 20: Case 2 COE

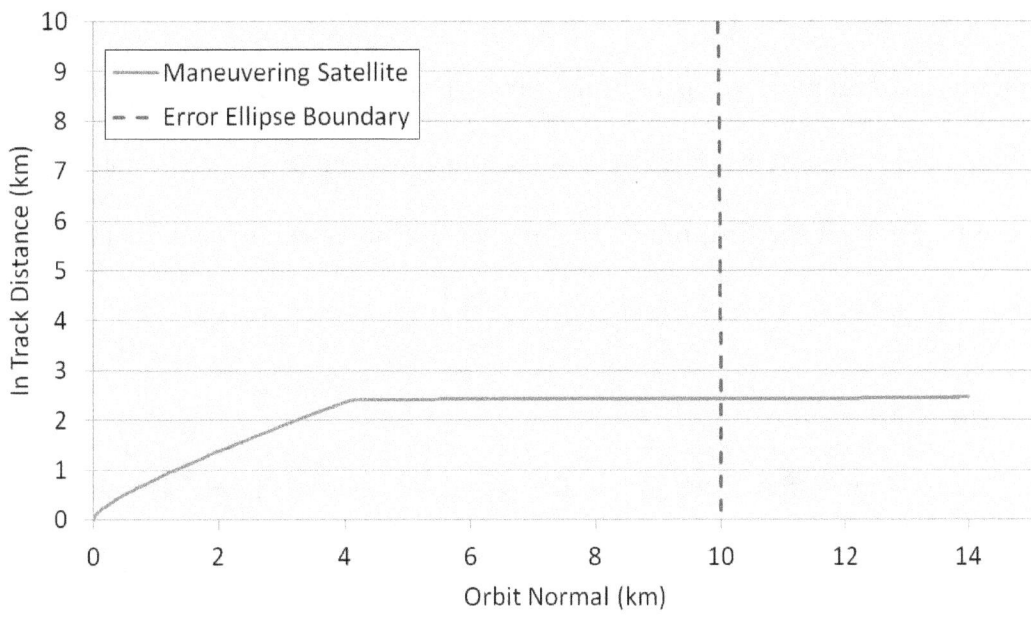

Figure 21: Case 2 Orbit Relative Motion Cross Section

out-of-plane nature, the thrusting angle shown in this figure is yaw rather than pitch. The resulting Two-Body orbital elements for this solution are given in Figure 20 and its cross sectional plot is given in Figure 21. For this case, the cross sections depict the orbit normal component relative to the in-track axis.

This solution demonstrates thrusting entirely out of the orbital plane in order to maneuver out of the ellipse. Thus, the timing of this impulse is as important as the duration. Figure 19 demonstrates placing the thrust a quarter of an orbit prior to AOR reentry. Another nearly identical maneuver may also be conducted three quarters of an orbit prior to re-entry without significantly affecting the cost or result. As this figure demonstrates, the thrust magnitude solution generated by GPOPS-II did not yield a constant maximum thrust. This fluctuation in thrust is due to the automatic scaling used in the design of the Optimal Control Problem combined with slight inaccuracies in the GPOPS-II solution.

From these results it can be seen that the resulting changes in altitude and eccentricity are negligible when thrusting out-of-plane. This case focuses instead on modifying the inclination and RAAN in order to achieve out-of-plane spacing from the reference trajectory at the final time. The estimated delta v requirement for this case is 14 m/s.

4.2.4   Case 2 Multiple Orbit

The multiple orbit solution for the Impulsive Out-of-Plane Case showed no significant advantages over the single orbit solution. While there is a very slight change in the semi-major axis for the maneuvering satellite, the drift caused by the difference in orbital periods is not significant over this scenario time and as such this solution still yields no maneuvering until a quarter orbit prior to AOR reentry followed by an identical maneuver to the single orbit scenario for this case.

4.2.5   Case 3 Single Orbit

For the Continuous Thrust single orbit scenario, the satellite was given the first set of initial conditions shown in Table 6 with the weighting factor set at 1×10^{-7}. The number of collocation points was initially set at 4 nodes per segments and was increased between 4 and 10 nodes per mesh iteration. The optimal thrust solution for this profile is given in Figure 22. The resulting Two-Body orbital elements for this solution are given in Figure 23 and the cross section is given in Figure 24.

This solution maneuvers the satellite to place apogee at the final position resulting in a similar final position to the Case 1 single orbit solution. Where the Impulsive Case controls perigee position by determining when to thrust, this case accomplishes the same

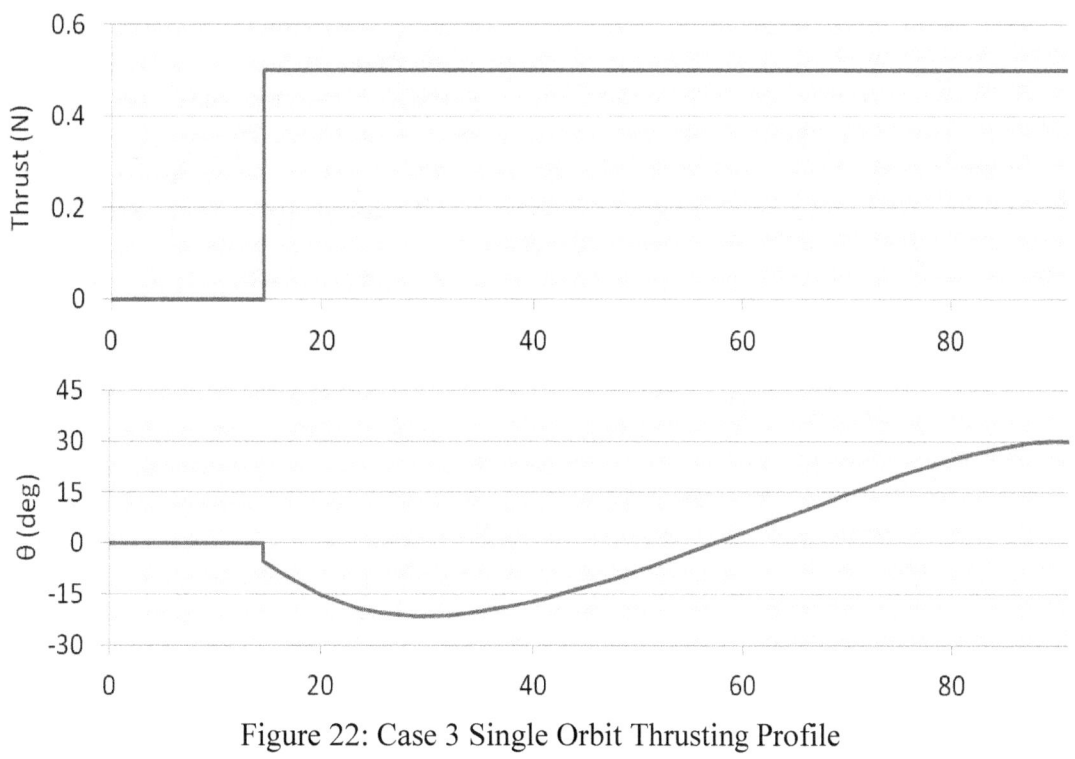

Figure 22: Case 3 Single Orbit Thrusting Profile

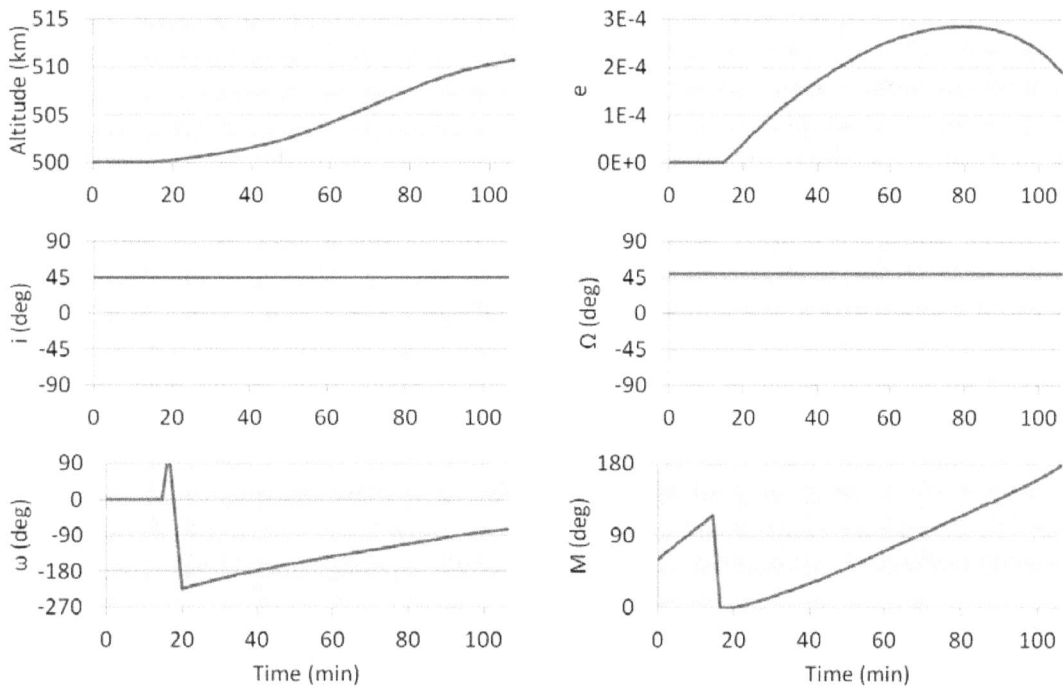

Figure 23: Case 3 Single Orbit COE

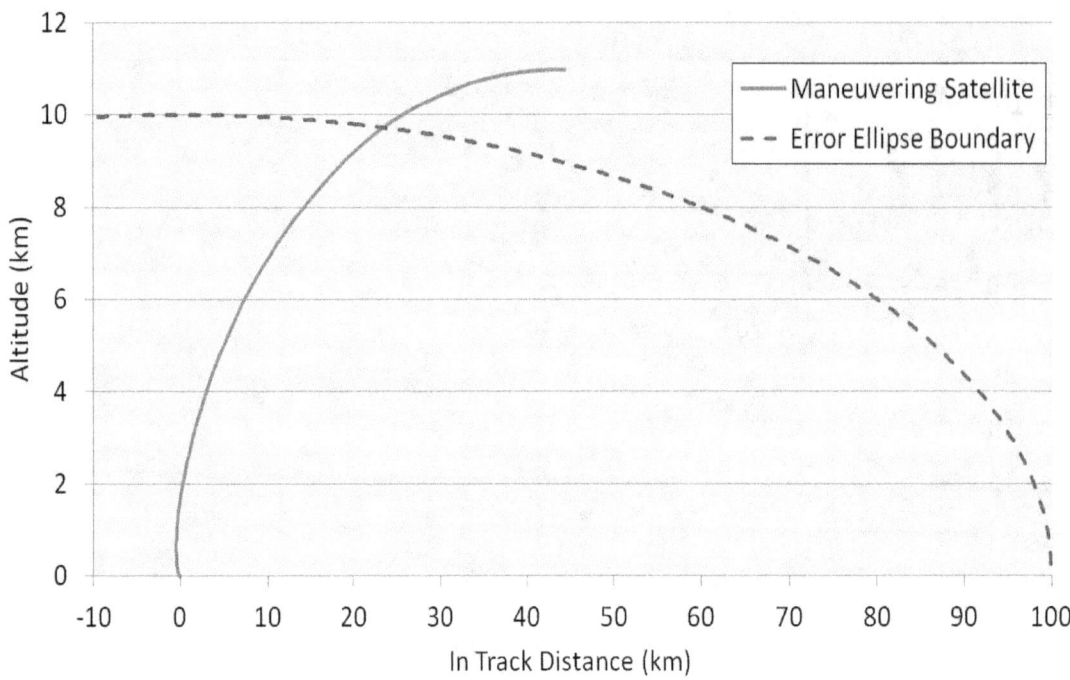

Figure 24: Case 3 Single Orbit Relative Motion Cross Section

goal using the pitch angle. The result is a solution that is not entirely in the velocity direction but rather oscillates within 30° of the velocity vector. It should be noted as well that since this case involves thrusting immediately after AOR departure, slightly larger in-track spacing is accomplished along with the altitude avoidance maneuver. This estimated maneuver cost was 5.5 m/s for this scenario.

4.2.6 Case 3 Multiple Orbit

For the Continuous Thrust multiple orbit scenario, the satellite was given the second set of initial conditions in Table 6 along with a weighting factor at 1×10^{-6}. The number of collocation points was initially set at 10 nodes per segment and was increased between 15 and 25 nodes per mesh iteration. The optimal thrust solution for this scenario is shown in Figure 25. The resulting Two-Body orbital elements for this solution are given in Figure 26 and the cross section for this maneuver is given in Figure 27.

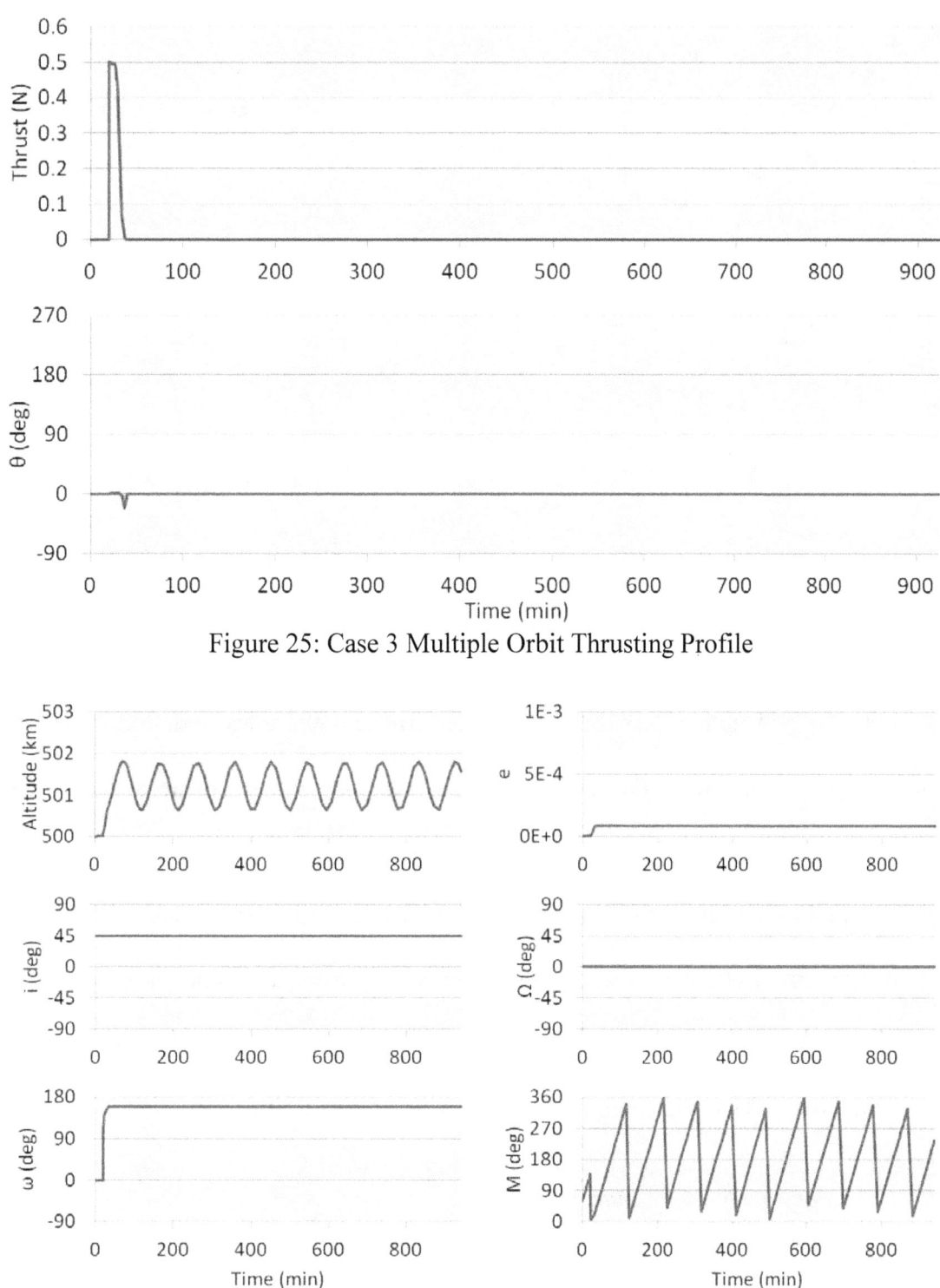

Figure 25: Case 3 Multiple Orbit Thrusting Profile

Figure 26: Case 3 Multiple Orbit COE

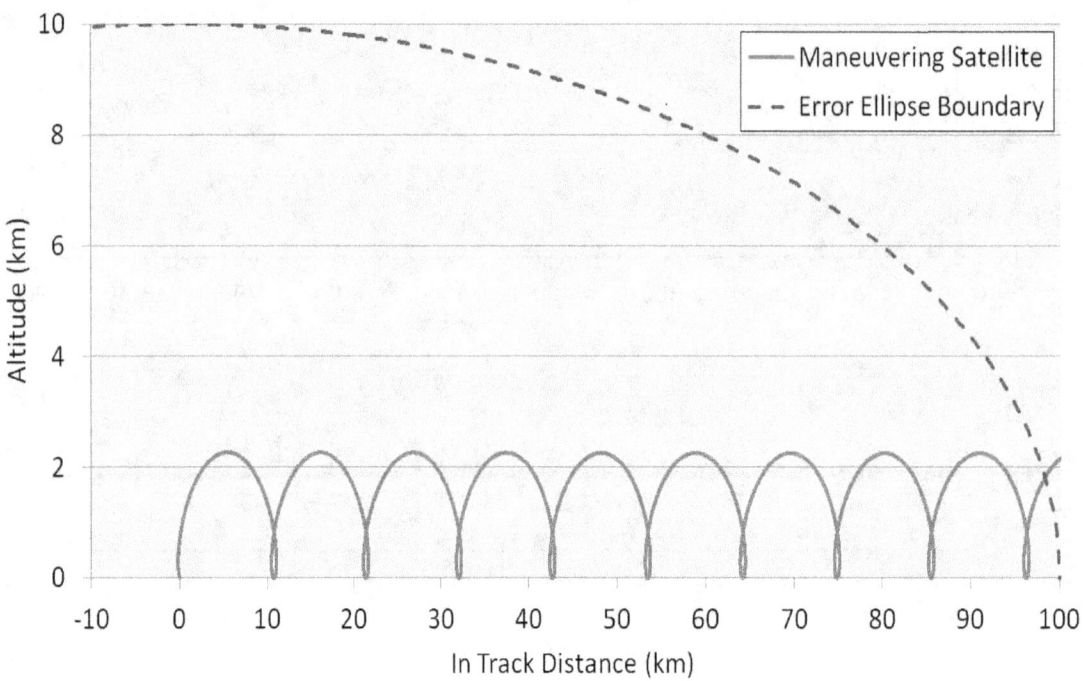

Figure 27: Case 3 Multiple Orbit Relative Motion Cross Section

Due to the length of this scenario, a much smaller delta v was required in order to achieve ellipse avoidance. For this reason, the resulting thrust profile appears more impulsive than it does continuous and represents a similar type of solution to the impulsive multiple orbit scenario, choosing to maneuver early for altitude and allowing the difference in orbital period to drive the increase in in-track distance in order to exit the ellipse. As with the Case 2 thrust profile, a slight deviation can be observed in the maximum thrust. The estimated maneuver cost was 0.7 m/s for this scenario.

4.2.7 Summary of Optimal Control Results

The Case 1 single orbit solution presents a viable alternative to the Hohmann Transfer, which would require maneuvering twice in order to re-circularize after changing altitudes. This solution focuses on something more closely related to a phasing maneuver, thrusting once and placing the furthest point from the reference orbit over the

AOR. In order to achieve ellipse avoidance using the Hohmann Transfer method for this scenario, a 37% smaller burn would be required initially upon AOR departure followed by an identical burn half an orbit later. The leads to a 26% increase in fuel costs to conduct the Hohmann Transfer over the single orbit solution for this case. The Case 1 multiple orbit solution allows for this single impulse to occur early in the profile, creating a slightly longer orbital period and allowing the new trajectory to diverge from its reference trajectory naturally. This maneuver requires far less fuel than the either the Hohmann Transfer or the phasing maneuver but unfortunately occurs with considerably less frequency than the single orbit scenario.

The Case 2 solution presents an interesting alternative to more traditional methods of maneuvering. Rather than attempting to change altitude or in-track spacing, this maneuver could be accomplished as late as a quarter orbit prior to AOR reentry such that the out-of-track spacing is maximized. This method provides for the most rapid response but unfortunately comes at the highest cost. This case alone shows no significant advantage in the multiple orbit scenario due to its negligible change in orbital period.

The Case 3 solution is similar in many respects to the solution to Case 1. It is by definition a more gradual change based on the nature of the engine being used. It should be noted that for the Case 3 single orbit solution a considerable amount of attitude maneuvering is required in order to accomplish the specified thrust vectoring for the single orbit scenario. The multiple orbit scenario is nearly identical to the Case 1 solution, requiring either velocity or anti-velocity thrusting for much shorter time periods than the single orbit scenario and allowing for the differences in orbital periods to generate maneuvering and reference trajectories that diverge.

The total fuel expenditures for each Case are given below in Table 7. For both the single orbit and the multiple orbit scenarios, the Impulsive In-Plane thrust solution yields the minimum delta v requirement while the Impulsive Out-of-Plane solution yields the maximum requirement.

Table 7: Fuel Cost Comparison

	Case		Δv (m/s)
1	Impulsive Thrust In Plane	Single Orbit	3.7
		Multiple Orbit	0.6
2	Impulsive Thrust Out-of-Plane	Single Orbit	14
3	Continuous Thrust	Single Orbit	5.5
		Multiple Orbit	0.7

4.3 Systems Tool Kit® Simulation and Validation

In addition to data generation, Systems Tool Kit® was also utilized in order to check the validity of the optimal solutions generated by GPOPS-II. This software also provided the ability not only to verify the Two-Body solutions but to also to demonstrate the effects that orbital perturbations have on the calculated maneuvers. The following sections provide the results when the profiles presented in Section 4.2 were implemented and propagated in STK using the full High-Precision Orbit Propagator (HPOP) engine.

4.3.1 Case 1

The simulation run for the Case 1 single orbit scenario yielded the results shown in Figure 28. The elements for the reference satellite are given in blue and represent the STK HPOP solution. The elements for the maneuvering satellite are given in red and also represent the HPOP solution. Since no out-of-plane thrusting was conducted, the inclination and RAAN were left out of this figure. Also, since the argument of perigee and mean anomaly are nearly identical for these two trajectories, those elements were also disregarded. It can be seen that the relative altitude changes occur as predicted by the GPOPS-II solution. The eccentricity plot also demonstrates that the orbit remains nearly circular within the bounds of normal perturbations. This deviation from the reference trajectory also changes predictably in accordance with the Two-Body approximation.

The results for the Case 1 single orbit pitch and yaw parameter searches are given below in Figure 29. The ellipse avoidance factor in this figure represents the value of the ellipse constraint as defined in Equation 51. The thrust angle parameter search yielded an optimal pitch angle at approximately 2° above the velocity direction for a climb and 2° below the anti-velocity direction for a descent. Both pitch angles had a corresponding yaw angle at zero. These results are nearly consistent with the solution from GPOPS-II presented previously in Figure 13 which indicated a 5° deviation from the velocity vector was optimal. The results from the Case 1 single orbit thrust duration parameter search are given below in Figure 30. This figure demonstrates that the thrust duration presented previously could have accomplished the ellipse avoidance with a delta v that was 6% smaller. This discrepancy is due to round off error in the conversion process within MATLAB® between the GPOPS-II output and STK. However, this deviation is on the order of 5 seconds and is well within the margin for error of a commanded maneuver. When the correction is made for this maneuver the delta v requirement becomes 3.5 m/s.

The simulation run for the Case 1 multiple orbit scenario yielded the results shown in Figure 31. As with the single orbit scenario, only altitude and eccentricity are presented in this plot. Despite the added perturbations in this figure, the differences in relative position between the two satellites remain consistent with GPOPS-II predictions.

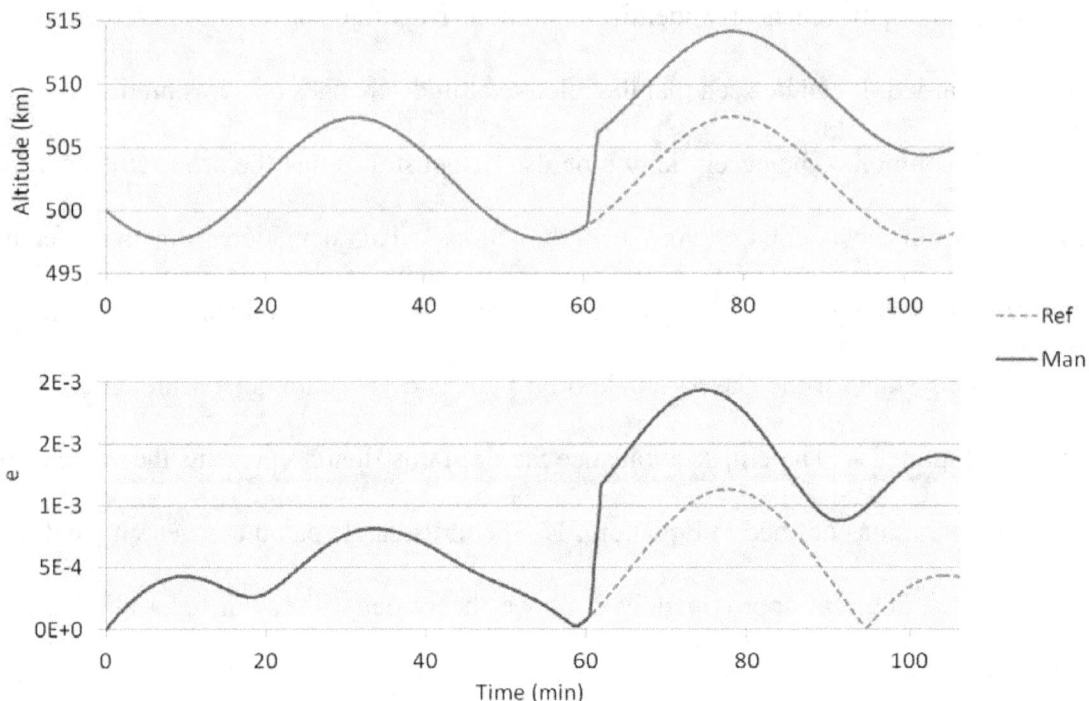

Figure 28: Case 1 Single Orbit STK Results

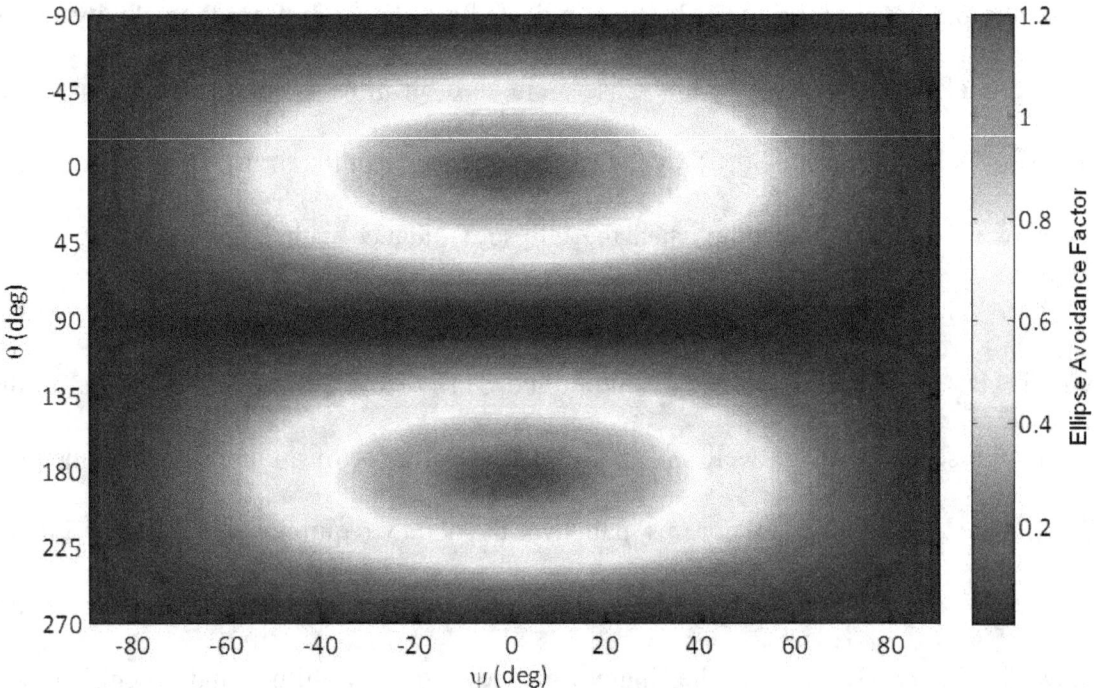

Figure 29: Case 1 Single Orbit Pitch and Yaw Validation

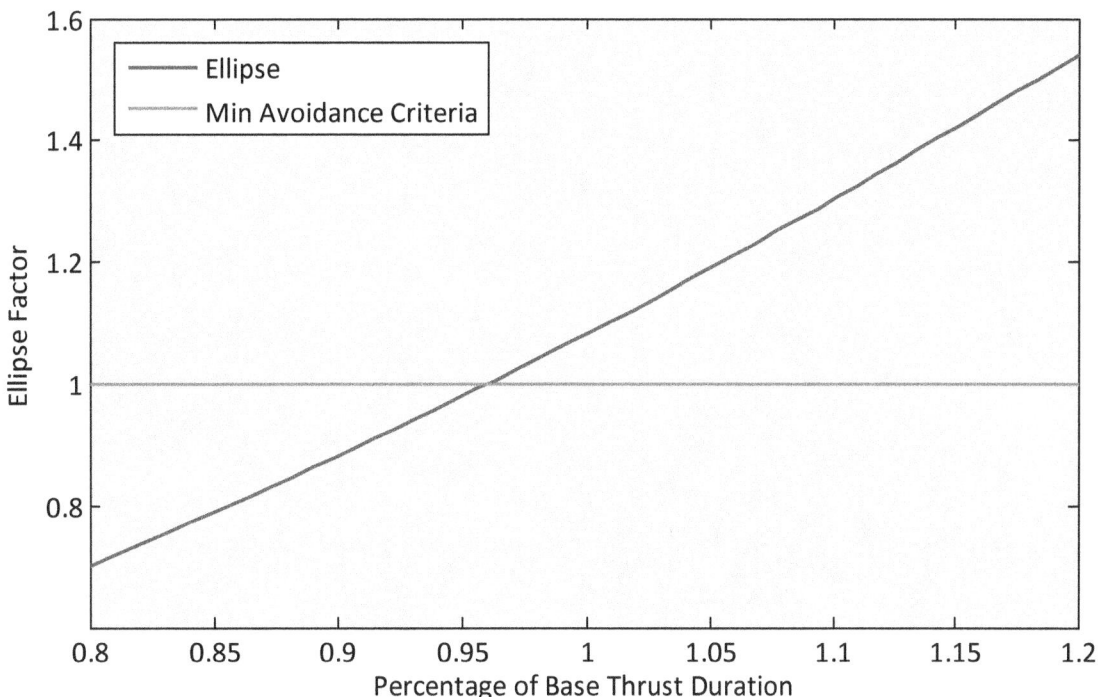

Figure 30: Case 1 Single Orbit Thrust Duration Validation

The results for the Case 1 multiple orbit pitch and yaw parameter searches are given in Figure 32. The thrust angle yielded an optimal pitch angle of 1.5° above the velocity vector for a climb and 1.5° below the anti-velocity vector for a descent. Both pitch angle solutions had a corresponding yaw angle at zero as in the single orbit scenario. These results are consistent with the profile presented previously in Figure 16. The results from the Case 1 multiple orbit thrust duration parameter search are given in Figure 33. This figure again demonstrates that the previously presented thrust duration could have been 13% smaller corresponding to a difference in thrust duration of 2 seconds and still accomplished the in track spacing necessary for ellipse avoidance. When this thrust duration is corrected, the delta v requirement for this maneuver becomes 0.5 m/s.

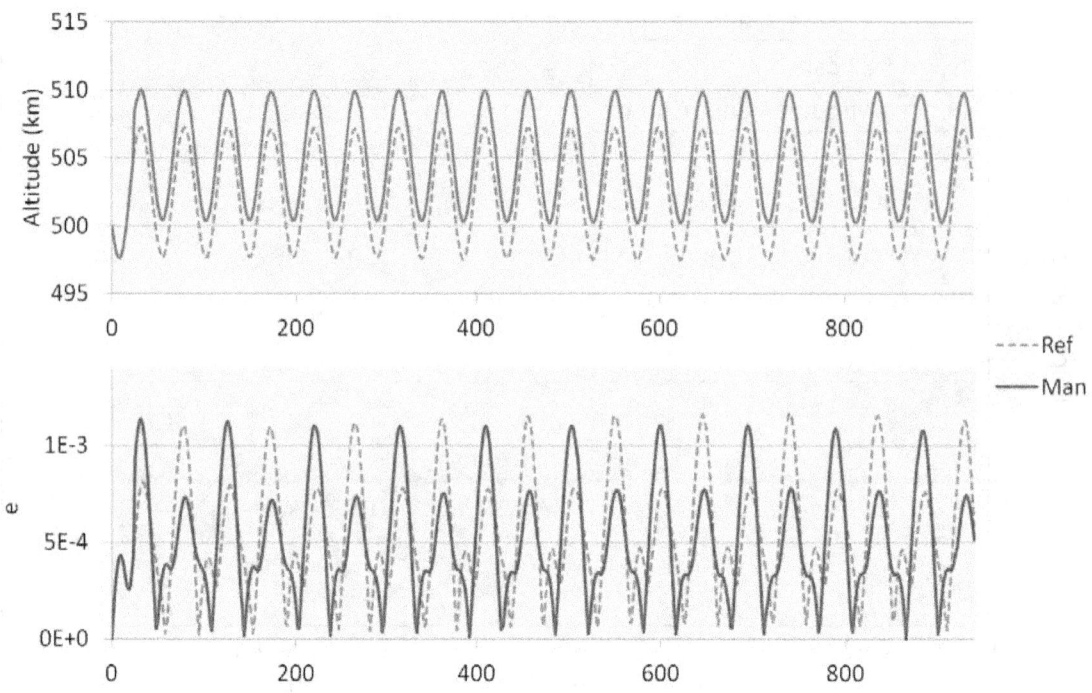

Figure 31: Case 1 Multiple Orbit STK Results

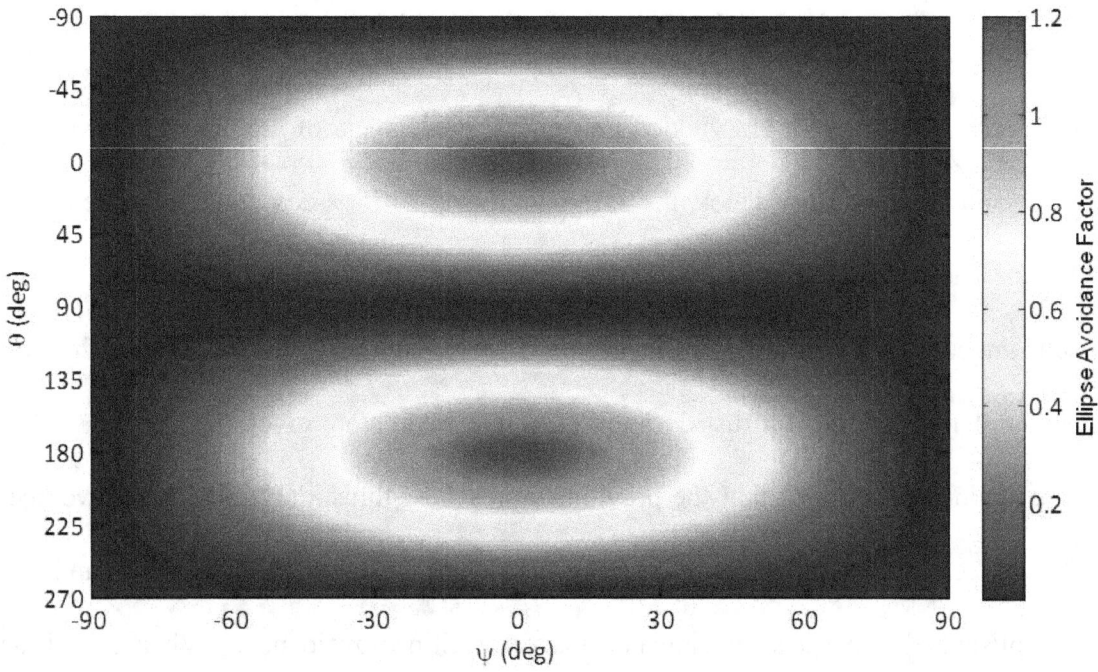

Figure 32: Case 1 Multiple Orbit Pitch and Yaw Validation

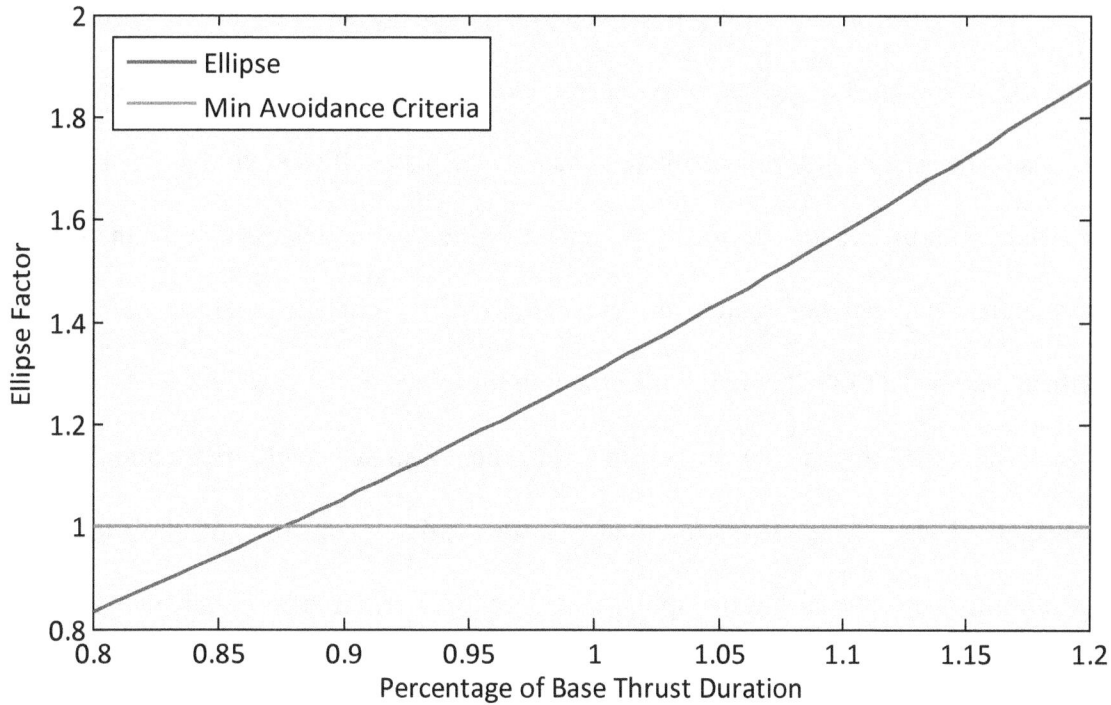

Figure 33: Case 1 Multiple Orbit Thrust Duration Validation

4.3.2 Case 2

The STK HPOP simulation results for Case 2 are shown below in Figure 34. Since this case does not significantly affect orbital period, the altitude, eccentricity, argument of perigee, and mean anomaly plots were excluded. The elements of interest shown for this case are inclination and RAAN. This figure shows responses in these elements consistent with their Two-Body predictions given in Section 4.2. It is interesting in this case to note that the changes made in inclination were on the order of the orbital perturbations while the changes in RAAN were an order of magnitude smaller than the perturbation effects. This would imply that this maneuver generates a negligible impact on the maneuvering satellite's mission effectiveness.

The results for the Case 2 pitch and yaw parameter searches are given below in Figure 35. The thrust angle parameter search indicated that pitch angle was irrelevant in this case since the only factor of interest was out-of-plane ellipse avoidance. This was consistent with the results previously presented in Figure 19. The results for the Case 2 thrust duration parameter search are given below in Figure 35. These two figures confirm that the GPOPS-II solution is optimal in this case.

The thrust magnitude from Figure 19 demonstrates slight fluctuations while thrusting. These fluctuations in thrust are an artificial construct of the GPOPS-II algorithm and are due to the manipulation of the code required to obtain this solution. However, the standard deviation of the thrust magnitude was 0.25 N and is well within the margin of error for a commanded maneuver.

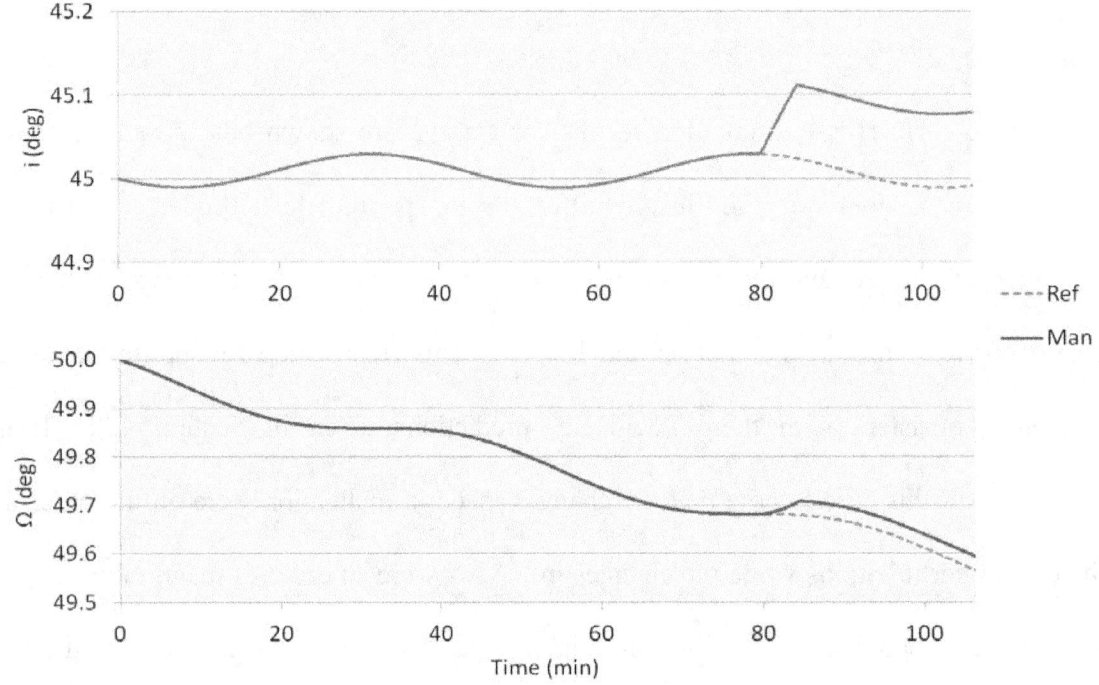

Figure 34: Case 2 STK Results

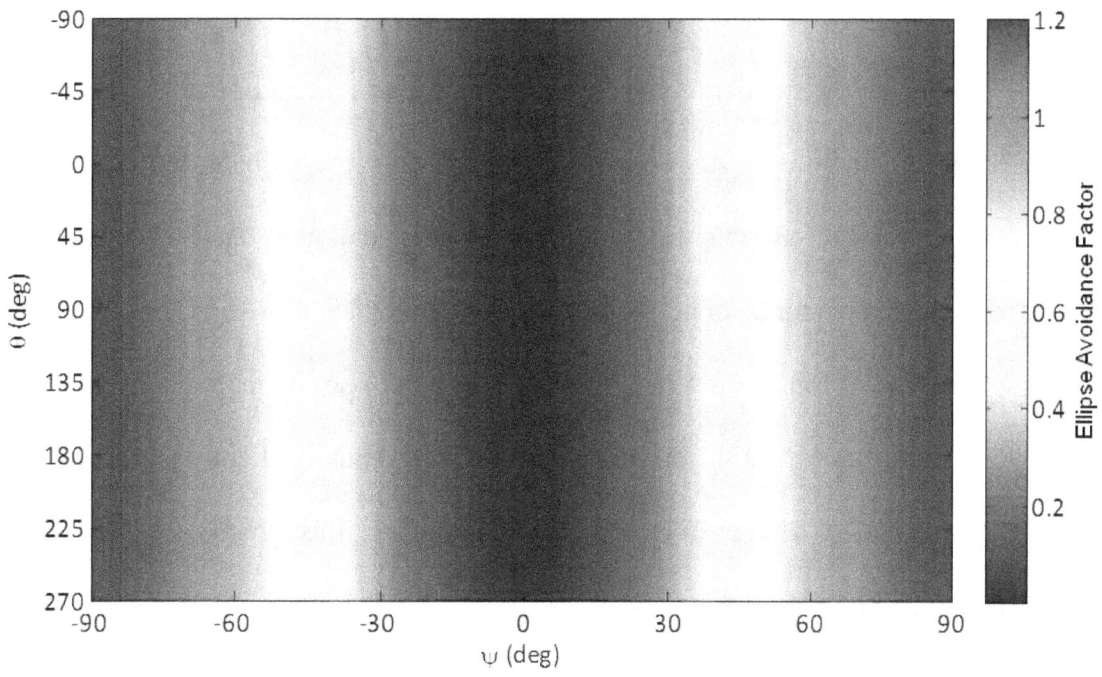

Figure 35: Case 2 Pitch and Yaw Validation

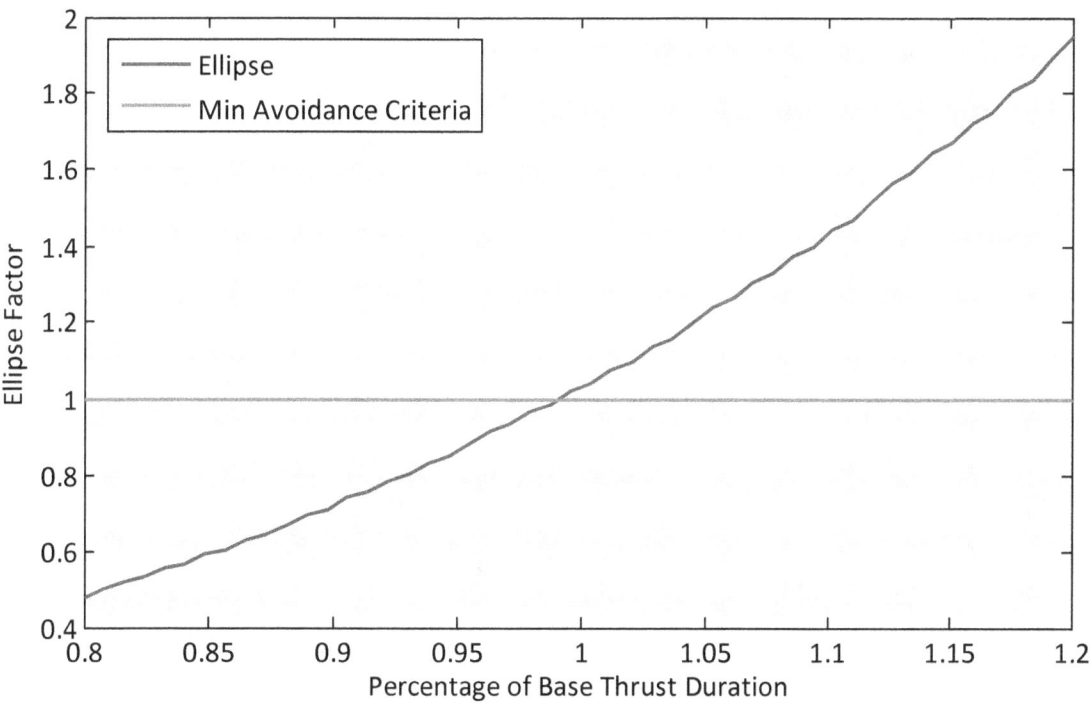

Figure 36: Case 2 Thrust Duration Validation

4.3.3 Case 3

The STK HPOP simulation for the Case 3 single orbit scenario yielded the results shown below in Figure 37. As with the first case, only the altitude and eccentricity values are shown. From this figure it can be seen that the relative altitude and eccentricity changes continue to be portrayed accurately by the Two-Body solution. The results for the Case 3 Single Orbit pitch profile parameter search are given in Figure 38. This figure demonstrates that the optimal pitch profile is actually slightly perturbed from the GPOPS-II solution. However, the optimal solution from this method has a maximum deviation from the GPOPS-II solution of 4° and is within a reasonable margin of error for a maneuvering satellite. Due to the fact that the single orbit scenario requires maneuvering for the entire scenario, no thrust duration validation was conducted.

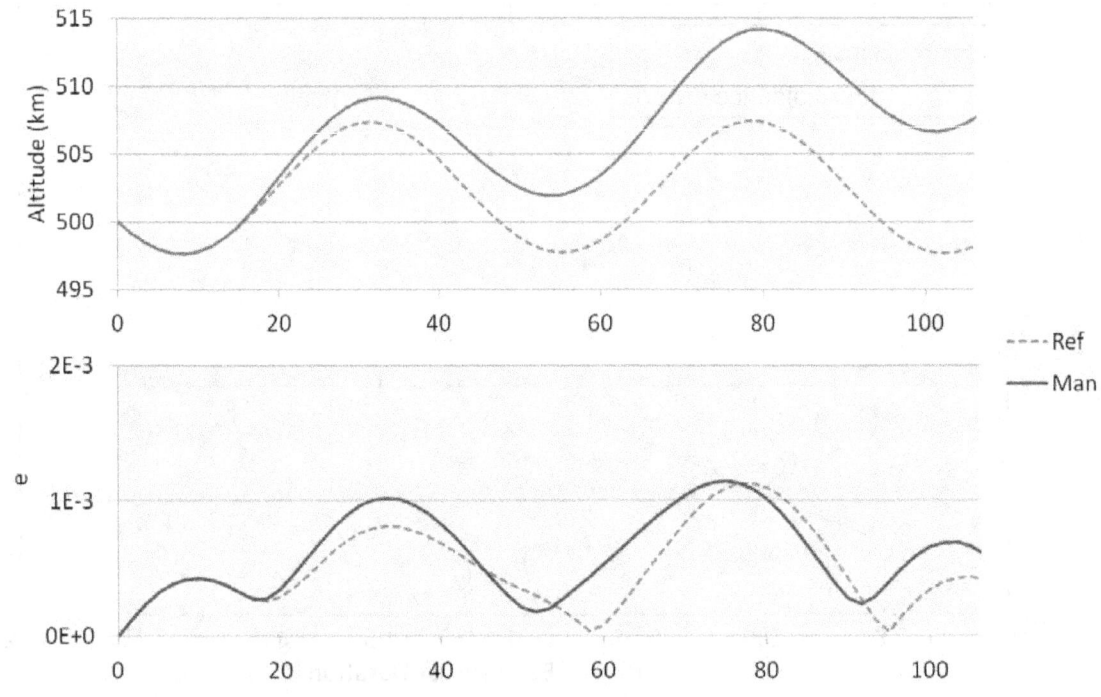

Figure 37: Case 3 Single Orbit STK Results

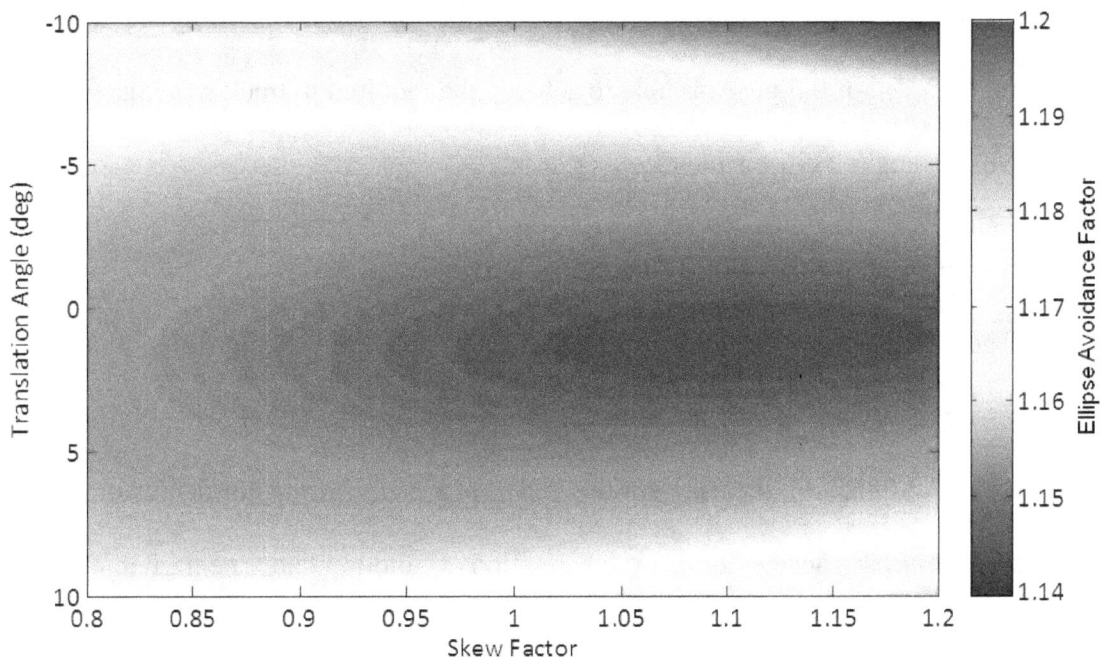

Figure 38: Case 3 Single Orbit Pitch Profile Validation

The STK HPOP simulation for the Case 3 multiple orbit scenario yielded the results shown below in Figure 39. As with the single orbit scenario only altitude and eccentricity are presented in this plot. Similar to the Case 1 multiple orbit scenario, the addition of perturbations still generate consistent differences in relative position between the two satellites with the GPOPS-II predictions.

The results for the Case 3 Multiple Orbit pitch and yaw parameter searches are given below in Figure 40. The thrust angle yielded an optimal pitch angle of 1.5 degrees for a climb and 1.5 degrees for a descent. The optimal yaw angle remained at zero as in Case 1. These results were consistent with the GPOPS-II solution for this case. The results for the Case 3 Multiple Orbit thrust duration parameter search are given in Figure 41. This figure indicates that the thrust duration determined by GPOPS-II is again slightly less than ideal. The validation routine returned an error of 13% and represents a difference of 91 seconds in thrust duration. When this deviation was corrected, a smaller delta v requirement of 0.5 m/s was able to achieve the required in track spacing necessary to achieve the required ellipse avoidance criteria.

4.3.4 Summary of STK Results

The STK simulation runs presented in this section demonstrate that these maneuvers will in fact create the changes predicted by the GPOPS-II solution from Section 4.2. Additionally, the usage of the HPOP engine in Astrogator demonstrates that the lack of perturbing accelerations in the equations of motion had a negligible effect on the calculation of valid final solutions.

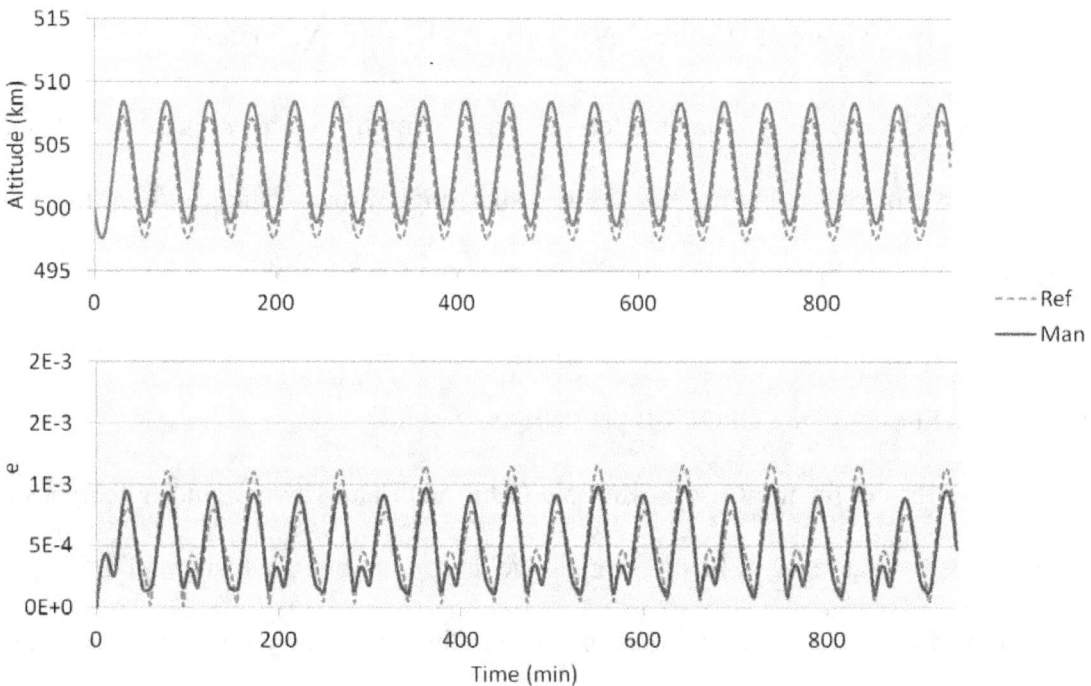

Figure 39: Case 3 Multiple Orbit STK Results

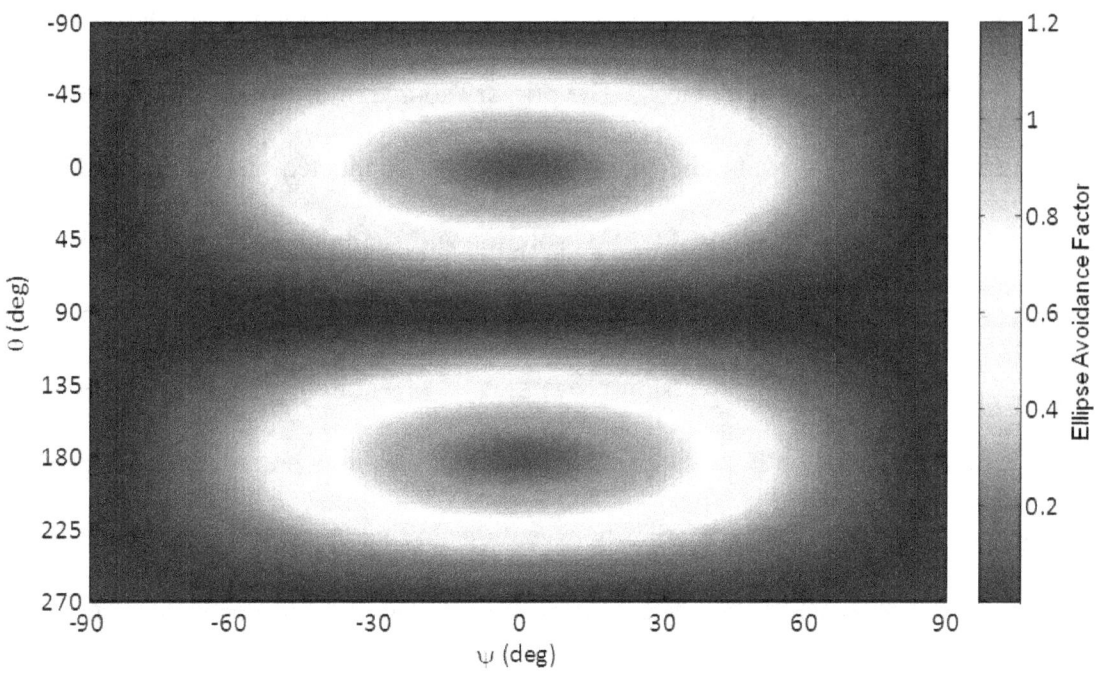

Figure 40: Case 3 Multiple Orbit Pitch and Yaw Validation

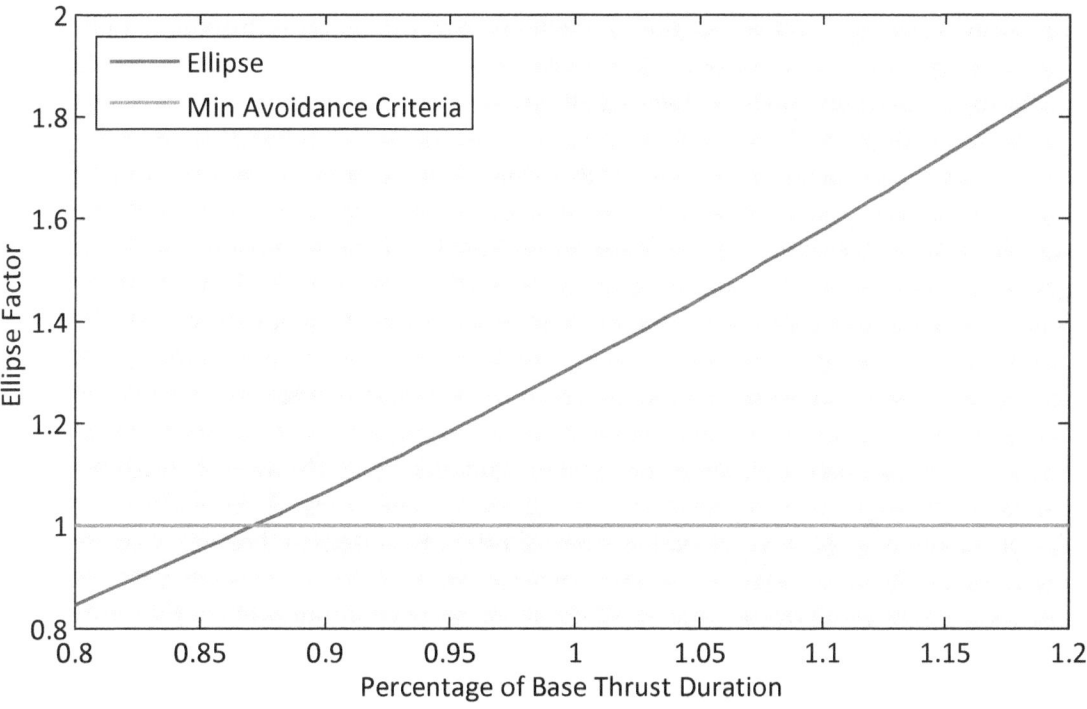

Figure 41: Case 3 Multiple Orbit Thrust Duration Validation

The parameter search yielded confirmation of the optimality of the GPOPS-II solution within reasonable tolerances. The primary source of error between the solutions presented in Section 4.2 and the validation routine conducted in Section 4.3 is the conversion process between the GPOPS-II solution and Astrogator within STK. Many of the thrust profile results from the GPOPS-II solution indicated a magnitude or angle that had small deviations. The conversion process to STK required the removal of many of these deviations. Even with these removals, the final solutions presented did not deviate significantly in most cases. The primary exceptions to this are the thrust duration results for each of the multiple maneuver cases. In both of these cases, GPOPS-II depicted a significantly larger thrust than was strictly required for ellipse avoidance.

Together, these three cases yielded three optimal families of solutions. First, if time permits it is most advantageous to make a small increase in altitude and allow time for the difference in orbital periods to slowly increase in-track spacing. If that is not possible, the next best solution involves thrusting in order to place apogee or perigee over the AOR reentry position. For the impulsive cases, this is accomplished using the timing of the thrust and for the In-Plane case involves a slightly lower delta v than the Continuous case which uses pitch angle to control apogee or perigee. This maneuver is roughly seven to nine times more expensive than the first solution. The least efficient solution involves making very small changes in the inclination and/or RAAN at a quarter or three quarters of an orbit prior to AOR reentry. This solution costs roughly three to four times the fuel cost of the single orbit solution and nearly twenty-four times the fuel cost of the multiple orbit solution. Its primary advantage is in maintaining previous altitude and eccentricity. Together, these three solutions outline the optimal set of potential maneuvers for collision avoidance.

4.4 Chapter Summary

This chapter presented the results from the optimal control problem solved using GPOPS-II in Section 4.2. Next, Section 4.3 demonstrated these maneuvers in STK and tested how adding perturbing accelerations altered the solution. STK was also utilized in this chapter to further optimize the solution and present reasonable minimum fuel requirements for each maneuver. The next chapter will present conclusions from this research and recommended future work.

V. Conclusions and Recommendations

5.1 Chapter Overview

The purpose of this research was to develop and test the application of pseudospectral optimization for debris avoidance in orbital mechanics. This study focused on the development of a set of minimum thrust maneuvers for the purpose of orbital debris collision avoidance. These thrust profiles were determined from the requirement that a satellite maneuver result in an orbit that is completely outside of an error ellipse of fixed dimensions projected from its non-thrusting reference trajectory within a set time frame. This was accomplished via GPOPS-II, a pseudospectral optimal control algorithm designed to run in MATLAB®. The results from this work were further developed and tested using the Component Object Model Interface to automate functionality in Systems Tool Kit® in order to propagate the calculated thrust profiles and compare the relative position between the maneuvering satellite and its reference trajectory.

This research developed maneuvers for three specific cases. The first case consisted of an impulsive thrust profile in the satellite's orbital plane. The second case maintained the impulsive nature while considering maneuvering independent of the satellite's orbital plane. The final case compared continuous thrusting to the impulsive case. STK was used for each of the three cases in order to validate the calculated optimal solutions as well as to demonstrate the effects of adding perturbations to the propagated trajectories.

5.2 *Conclusions*

Both the problem setup and solution developed in this study demonstrated the viability of GPOPS-II as an optimal control algorithm for application in orbital mechanics as well as serving as a basis for future study in this area. However, the sensitivity of this algorithm to scenario settings indicates that this tool is best suited for theoretical maneuver development in a controlled environment. Small changes to the scenario settings within this problem required extensive manipulation of variables such as the cost function weighting factor and nodal distribution in order to obtain meaningful results.

The results from the three test cases demonstrated that the most efficient way to maneuver out of the error ellipse consisted of thrusting mostly in the velocity or anti-velocity direction with a single impulse. If time permits, it is most efficient to thrust for a shorter time with the intent of slightly changing the orbital period. This allows the maneuvering satellite to slowly diverge from its reference trajectory, allowing for separation dependent almost entirely on in-track spacing to maneuver out of the ellipse. If time does not permit the in-track solution, however, the next best option consists of maneuvering to place apogee or perigee over the final position. This method allows for the satellite to leave the ellipse temporarily for the collision avoidance maneuver but does not attempt to re-circularize the orbit afterwards. As expected, out-of-plane thrusting was shown to be the least efficient but had the advantage of an almost negligible change in virtually all of the orbital elements.

Analysis in STK demonstrated the effects of the addition of perturbations into the propagator after convergence of the optimal control algorithm. This analysis demonstrated that typical perturbing forces did not significantly change the predicted trajectories of the maneuvering satellites relative to their non-maneuvering reference trajectories. This was consistent with the initial assumption that only Two-Body mechanics were necessary to properly model the dynamics of this problem.

5.3 Research Limitations

Every optimal control problem begins with the question of what, exactly, constitutes optimality. This is specified in the problem statement in the form of the cost function. This research made use of specific choices for several values used in the cost function. Obviously, there are a nearly infinite number of possible permutations of these choices available for even this single formulation of the cost function, not even including additional forms designed to alternately express either the Mayer error ellipse penalty or the Lagrange minimum fuel running cost. Therefore, the claim that these trajectories are optimal or even near-optimal is made only after test runs were conducted in STK to verify functionality and optimality of the solution. Different problem formulations could potentially yield better solutions in terms of optimality and robustness.

5.4 Recommendations for Future Work

The STK Component Object Model Interface library developed in this research was designed broadly with the intention of providing an automation tool for future research requiring rapid communication and control of STK from within MATLAB®. While this research utilized this tool to facilitate optimization in orbital maneuvering, autonomous control from MATLAB® yields a wide variety of data processing and scenario generation options not currently available in STK by itself. Further development of this tool to expand functionality would be extremely beneficial for future study in orbital mechanics.

This research was conducted using pseudospectral optimization to determine a set of appropriate thrust profiles for collision avoidance. Future work should include an analytical approach, such as Primer Vector Theory, that could be used to further validate the methods presented in this study. An analytical approach would offer the ability to study how alterations to this scenario such as satellite mass and maximum thrust would affect the solution. Additionally, the use of alternate direct optimization routines as well as alternate problem formulations would be advantageous in order to compare accuracy and convergence times. Alternate problem formulations should include techniques to automatically scale the weighting factor in the cost function. Static values for this weighting factor provided one of the primary limitations in the robustness of the algorithm developed in this research. Further development of the multiple phase formulation should also be explored to better model impulsive thrusting.

Another potential area of future study for this research would be to analyze the effects, if any, that these maneuvers would have on a constellation of satellites. Maintaining relative positions is critical to a properly functioning satellite constellation. While the maneuvers covered in this research are by design extremely small, their effects on a constellation of satellites might still degrade overall coverage and should therefore be explored. This would require expansion into perturbation theory within the dynamics of the optimal control problem since the proximity assumption used in this research is no longer valid.

This research briefly considered how general perturbing accelerations affected the difference in relative position from the Two-Body solution and demonstrated that these effects had a negligible impact on the ellipse avoidance. However, it is conceivable that incorporating perturbation effects into the equations of motion prior to the optimization step might allow the satellite to use these effects to further improve maneuver efficiency and should be considered as an additional area for future study.

Appendix A. MATLAB® Code

The MATLAB® code used in this research merges the use of GPOPS-II to solve the optimal control problem via the Radau Pseudospectral Method with the visualization and propagation capabilities from Systems Tool Kit®. STK is used both as a visualization tool as well as a source of realistic data input. This code is broken out into the main code and two structures of functions. The first structure is the RO structure and contains the function library used in the main code in order to set up and run GPOPS-II. The second structure is the STK Component Object Model Interface Library designed to facilitate automatic communication between STK and MATLAB®.

A.1 Responsive Orbits Main Code

```
%% Created by James Sales
clear all; close all; clc;
global Scen
%% Select Thesis Case to Run
fprintf(1,'Please select a case:\n');
fprintf(1,['\t 1: Impulsive In Plane Single Orbit\n']);
fprintf(1,['\t 2: Impulsive In Plane Multiple Orbit\n']);
fprintf(1,['\t 3: Impulsive Out of Plane\n']);
fprintf(1,['\t 4: Continuous Single Orbit\n']);
fprintf(1,['\t 5: Continuous Multiple Orbit\n']);
p2              = input('>> ');
Scen.InPlane    = 1;
Scen.Continuous = 0;
switch p2
    case 1
        Scen.T_max    = 22/1e3;
        omega         = 50;
        Scen.Fraction = 0.1;
        Scen.alpha    = 9e-3;
        Scen.Nodes    = [20 25 7];
        Scen.angle    = 0;
    case 2
        Scen.T_max    = 22/1e3;
        omega         = 0;
        Scen.Fraction = 0.05;
```

```
            Scen.alpha     = 1e-2;
            Scen.Nodes     = [20 25 25];
            Scen.angle     = 0;
        case 3
            Scen.InPlane   = 0;
            Scen.T_max     = 22/1e3;
            omega          = 50;
            Scen.Fraction  = 1;
            Scen.alpha     = 1e-4;
            Scen.Nodes     = [4 10 4];
            Scen.angle     = pi/2;
        case 4
            Scen.T_max     = 0.5/1e3;
            omega          = 50;
            Scen.Fraction  = 1;
            Scen.alpha     = 1e-7;
            Scen.Nodes     = [4 10 4];
            Scen.angle     = 0;
            Scen.Continuous = 1;
        case 5
            Scen.T_max     = 0.5/1e3;
            omega          = 0;
            Scen.Fraction  = 0.1;
            Scen.alpha     = 1e-6;
            Scen.Nodes     = [15 25 10];
            Scen.angle     = 0;
            Scen.Continuous = 0;
        otherwise
            fprintf(1,'\n Error: Incorrect entry.  Please try again.\n');
            return
end
%% Set Commonly Manipulated Variables
Scen.NumDays    = 10;
Scen.Prop       = 'Earth HPOP Default v8-1-1';
Scen.m_sat      = 400;
Scen.m_fuel     = 100;
Scen.Path       = 'I:\My Documents\Thesis\STK Test Runs\';
% Satellite IC's:    [a      e      i      omega    w    M ]
Scen.COE        =    [6878   1e-6   45     omega    0    60];
% Specify Area Target dimmensions and location for the Midwest Scen
Scen.Centroid   = [39.7828, 275.917, 0];
Scen.Size       = [500, 500, 0];
Scen.ElevAngle  = 20;                                       % deg
Scen.mu         = 3.98601e5;                                % km^3/s^2
Scen.Re_e       = 6378;                                     % km
% Convert COE's to Equinoctal Elements      [a h k p q F]
```

```
Scen.EE(1)       = Scen.COE(1);
Scen.EE(2)       = Scen.COE(2)*sind(Scen.COE(4)+Scen.COE(5));
Scen.EE(3)       = Scen.COE(2)*cosd(Scen.COE(4)+Scen.COE(5));
Scen.EE(4)       = tand(Scen.COE(3)/2)*sind(Scen.COE(4));
Scen.EE(5)       = tand(Scen.COE(3)/2)*cosd(Scen.COE(4));
Scen.EE(6)       = (Scen.COE(4)+Scen.COE(5)+Scen.COE(6))*pi/180;
%% Set Start and End Times and format for use in STK
Scen.clock       = [2013, 1, 1, 9, 0, 0];
Scen.Now         = 0;
Scen.StartTime   = RO.Time_Sequencer(Scen.clock, 0);
Scen.EndTime     = RO.Time_Sequencer(Scen.clock, Scen.NumDays*86400);
% Create unique title based on current date and time
if Scen.StartTime(3)==' '
    Scen.Title   = [Scen.StartTime(1:2) Scen.StartTime(4:6) Scen.StartTime(8:11)];
else
    Scen.Title   = [Scen.StartTime(1) Scen.StartTime(3:5) Scen.StartTime(7:10)];
end
Scen.Epoch       = Scen.StartTime;
Scen.TimeStep    = 10;
%% Initialize STK and create Scen components
[uiapp, root]    = STK.Initialize(Scen);
[ref, MCS_r]     = STK.Astrogator('R',root,Scen);
target           = STK.Area_Target('AOR',root,Scen);
% Create Engine to meet specs listed above
Scen.EngineName = 'Responsive Orbits Engine Model';
STK.Create_Engine_Model(root,Scen.EngineName, Scen.T_max*1e3);
% Set the Reference satellite to propagate for 1 day
STK.Propagate('Reference Trajectory',86400*10,MCS_r, Scen.Prop);
ref.Graphics.Attributes.Intervals.RemoveAll;
ref.Graphics.Attributes.Default.Inherit = 0;
ref.Graphics.Attributes.Default.IsOrbitVisible = 0;
ref.Propagator.RunMCS;
%% Compute access times and determine coast and maneuvering profile durations
Scen.AccessTimes = STK.Compute_Access(root,ref,target,Scen.clock);
Scen.Coast       = Scen.AccessTimes.EpSec(1,2);
Scen.t           = Scen.AccessTimes.EpSec(2,1)-Scen.Coast;
%% Account for coast time before entering data into GPOPS
Out.coast_t      = linspace(Scen.Now,Scen.Now+Scen.Coast,15);
[Out.t,Out.z]    = ode45(@RO.ODE_dynamics,Out.coast_t,Scen.EE);
Scen.EE          = Out.z(end,:);
%% Run GPOPS and retrieve Optimal Profile Solution
Solution         = RO.Run_GPOPS();
% Convert states out of GPOPS solution into COE's
Solution.phase.time = Solution.phase.time + Scen.Coast;
Out.length       = length(Out.t);
Out.t            = [Out.t; Solution.phase.time];
```

```matlab
% solution.state    = [a h k p q F]
Out.a              = [Out.z(:,1); Solution.phase.state(:,1)];
Out.h              = [Out.z(:,2); Solution.phase.state(:,2)];
Out.k              = [Out.z(:,3); Solution.phase.state(:,3)];
Out.p              = [Out.z(:,4); Solution.phase.state(:,4)];
Out.q              = [Out.z(:,5); Solution.phase.state(:,5)];
Out.F              = [Out.z(:,6); Solution.phase.state(:,6)];
Out.e              = sqrt(Out.h.^2+Out.k.^2);
Out.i              = 2.*atan(sqrt(Out.p.^2+Out.q.^2));
Out.omega          = atan2(Out.p,Out.q);
Out.w              = atan2(Out.h,Out.k)-atan2(Out.p,Out.q);
Out.M              = Out.F-atan2(Out.h,Out.k);
for count=1:length(Out.M)
    while Out.M(count)>2*pi
        Out.M(count) = Out.M(count) - 2*pi;
    end
end
% Read controls out of GPOPS Out structure
Out.T              = [zeros(length(Out.z),1);Solution.phase.control(:,1)]...
                        *Scen.T_max*1e3;                              % N
Out.Thrusting      = [];
for count = 1:length(Out.T)
    if Out.T(count)          < 1e-3
        Out.theta(count,1) = 0;
        Out.psi(count,1)   = 0;
    else
        Out.theta(count,1) = Solution.phase.control(count-length(Out.z),2);
        Out.psi(count,1)   = Solution.phase.control(count-length(Out.z),3);
        Out.Thrusting      = [Out.Thrusting;
                              Out.t(count) Out.theta(count) Out.psi(count)];
    end
end
%% Convert controls into ECI Componants and write to text file
Out.ECI            = RO.Convert_to_ECI(Out,Scen);
Scen.Dur           = STK.Output_to_text(Scen, Out, Out.length, 1);
%% Set the Maneuvering satellite to respond to the calculated trajectory and propagate
[man, MCS_m]       = STK.Astrogator('M',root,Scen);
if Scen.Continuous == 1
    STK.Propagate('Coasting',Out.coast_t(end),MCS_m,Scen.Prop);
    STK.Maneuver_From_File('GPOPS_Profile',MCS_m,Scen,1);
else
    STK.Propagate('Coasting',Out.Thrusting(1,1),MCS_m,Scen.Prop);
    v              = [cos(Out.Thrusting(1,2))*cos(Out.Thrusting(1,3));
                      sin(Out.Thrusting(1,3));
                      sin(Out.Thrusting(1,2))*cos(Out.Thrusting(1,3))]';
    STK.FTV_Maneuver('GPOPS_Profile', MCS_m, v, Solution.phase.integral);
```

```
end
STK.Propagate('Propagate',86400,MCS_m,Scen.Prop);
man.Graphics.Attributes.Intervals.RemoveAll;
man.Graphics.Attributes.Default.Inherit = 0;
man.Graphics.Attributes.Default.IsOrbitVisible = 0;
man.Propagator.RunMCS;
%% Plot Data in MATLAB
RO.XLSWrite();
Out.dv = Solution.phase.integral*Scen.T_max*1e3/(Scen.m_sat + Scen.m_fuel);
fprintf(1,'Total Delta v for the maneuver shown is: %4.1f m/s \n',Out.dv);
```

A.2 Responsive Orbits Function Library

```
classdef RO
% Created by James Sales
% Establishes the function library for the Responsive Orbits main code.
properties
end
methods(Static)
    function[Time] = Time_Sequencer(clock, Now)
        % Takes a MATLAB-standard clock vector as input along with the
        % variable 'Now' in seconds.  This function is used in the main
        % code in order to convert MATLAB clock time to an STK-compatible
        % input.
        
        % Break Scen.Now down into ellapsed days, hours, minutes, & seconds
        Days        = floor(Now/86400);
        Hours       = floor((Now-86400*Days)/3600);
        Minutes     = floor((Now-86400*Days-3600*Hours)/60);
        Seconds     = floor((Now-86400*Days-3600*Hours-60*Minutes));
        Month_str   = ['Jan';
                       'Feb';
                       'Mar';
                       'Apr';
                       'May';
                       'Jun';
                       'Jul';
                       'Aug';
                       'Sep';
                       'Oct';
                       'Nov';
                       'Dec'];
        if round(clock(1)/4) == clock(1)/4
            DPM         = [31;29;31;30;31;30;31;31;30;31;30;31];
        else
            DPM         = [31;28;31;30;31;30;31;31;30;31;30;31];
```

```
end
% Wrap Seconds, Minutes, Hours, Days, Months to make a legible date.
if clock(6) + Seconds >= 60
    clock(6) = clock(6) + Seconds - 60;
    Minutes          = Minutes + 1;
else
    clock(6) = clock(6) + Seconds;
end
if clock(5) + Minutes >= 60
    clock(5) = clock(5) + Minutes - 60;
    Hours            = Hours + 1;
else
    clock(5) = clock(5) + Minutes;
end
if clock(4) + Hours >= 24
    clock(4) = clock(4) + Hours - 24;
    Days             = Days + 1;
else
    clock(4) = clock(4) + Hours;
end
if clock(3) + Days > DPM(clock(2))
    clock(3) = clock(3) + Days - DPM(clock(2))+1;
    clock(2) = clock(2) + 1;
else
    clock(3) = clock(3) + Days;
end
if clock(2) > 12
    clock(2) = 1;
    clock(1) = clock(1)+1;
end
if clock(3)<10
    Day = ['0',num2str(clock(3))];
else
    Day = num2str(clock(3));
end
Today     = [Day,' ',Month_str(clock(2),:),' ',num2str(clock(1))];
if clock(4)<10;
    Hour = ['0',num2str(clock(4))];
else
    Hour = num2str(clock(4));
end
if clock(5)<10;
    Min = ['0',num2str(clock(5))];
else
    Min = num2str(clock(5));
end
```

```matlab
        if clock(6)<10;
            Sec = ['0',num2str(clock(6))];
        else
            Sec = num2str(clock(6));
        end
        Time    = [Hour,':',Min,':',Sec];
        Time    = [Today,' ',Time];
end

function [zd] = ODE_dynamics(t,z)
    % Non-maneuvering equations of motion for ODE 45.  This function is
    % used in the main code in order to model the coast time prior to
    % AOR departure in STK as well as to forecast the reference
    % satellite position in GPOPS.
    
    %% Define constants
    % Defined in Responsive_Orbits
    global Scen
    %% State and control Vector Inputs
    % EOM are computed in Equinoctal Elements    [a h k p q F]
    a       = z(1);
    n       = sqrt(Scen.mu/a^3);
    % State Derivatives
    zd(1) = 0;
    zd(2) = 0;
    zd(3) = 0;
    zd(4) = 0;
    zd(5) = 0;
    zd(6) = n;
    % ode45 requires column vectors as output
    zd=zd';
end

function [Solution] = Run_GPOPS()
    % This file builds the GPOPS-II input structure.  It delineates
    % state, control, and time limitations as well as providing an
    % appropriate guess.  It allows the main code to dictate the
    % different number of collocation points required for each scenario
    % being executed.
    
    %% Define constants
    global Scen REF
    t               = Scen.t;
    COE             = Scen.COE;
    %% Create Initial State Vector
    a               = Scen.EE(1);
```

```
h               = Scen.EE(2);
k               = Scen.EE(3);
p               = Scen.EE(4);
q               = Scen.EE(5);
F               = Scen.EE(6);
REF.z0          = [a h k p q F]';
%% Create Final State Vector for the REF satellite
% This utilizes ode45 to extrapolate the position of the non-manuevering
% satellite at the final time.
time            = linspace(0,t);
[time,zref]     = ode45(@RO.ODE_dynamics,time,REF.z0);
[row column]    = size(zref);
REF.zf    = zref(row,:)';
%% Determine reference satellite final position
a               = REF.zf(1);
h               = REF.zf(2);
k               = REF.zf(3);
p               = REF.zf(4);
q               = REF.zf(5);
F               = REF.zf(6);
root            = sqrt(1-h.^2-k.^2);
n               = sqrt(Scen.mu/a^3);
r               = a*(1-k*cos(F)-h*sin(F));
B               = 1/(1+root);
x               = a*((1-h^2*B)*cos(F)+h*k*B*sin(F)-k);
y               = a*(h*k*B*cos(F)+(1-k^2*B)*sin(F)-h);
% Determine Rotation Matrix R ir
i               = 2.*atan(sqrt(p.^2+q.^2));
REF.p           = p;
REF.q           = q;
REF.phi         = atan2(y,x);
REF.N           = x*cos(REF.phi)+y*sin(REF.phi);
REF.T           = y*cos(REF.phi)-x*sin(REF.phi);
%% Create bounds sub-structure for GPOPS
% State Limitations
bounds.phase.initialtime.lower  = 0;
bounds.phase.initialtime.upper  = 0;
bounds.phase.finaltime.lower    = t;
bounds.phase.finaltime.upper    = t;
bounds.phase.initialstate.lower = REF.z0;
bounds.phase.initialstate.upper = REF.z0;
bounds.phase.state.lower = [REF.z0(1)-5,-0.5,-0.5,-2*tan(i/2),-2*tan(i/2), -pi];
bounds.phase.state.upper = [REF.z0(1)+20, 0.5, 0.5, 2*tan(i/2), 2*tan(i/2),F+pi];
bounds.phase.finalstate.lower=[REF.z0(1),-0.5,-0.5,-2*tan(i/2),-2*tan(i/2),F-pi];
bounds.phase.finalstate.upper=[REF.z0(1)+20,0.5,0.5,2*tan(i/2),2*tan(i/2),F+pi];
bounds.phase.control.lower      = [0, -pi/2,-pi/2];
```

```matlab
    bounds.phase.control.upper    = [1,3*pi/2, pi/2];
    bounds.phase.integral.lower   = 0;
    bounds.phase.integral.upper   = t;
    %% Create guess sub-structure for GPOPS
    guess.phase.time     = time;
    guess.phase.state    = zref;
    n                    = round(Scen.Fraction*length(time));
    m                    = length(time) - n;
    guess.phase.control  = [[ones(1,n),zeros(1,m)]',zeros(m+n,1), ...
                           Scen.angle.*ones(m+n,1)];
    guess.phase.integral = Scen.Fraction*t;
    %% Build HP-adaptive mesh settings
    mesh.method = 'hp1';
    mesh.tolerance = 1e-8;
    mesh.maxiteration = 45;
    mesh.colpointsmin = Scen.Nodes(1);
    mesh.colpointsmax = Scen.Nodes(2);
    mesh.phase.colpoints = Scen.Nodes(3)*ones(1,10);
    mesh.phase.fraction =  0.1*ones(1,10);
    %% Concatenate substructures into setup input structure and run GPOPS
    setup.name = 'Responsive Orbits';
    setup.functions.continuous = @RO.Continuous;
    setup.functions.endpoint = @RO.Endpoint;
    setup.bounds = bounds;
    setup.guess = guess;
    setup.mesh  = mesh;
    setup.nlp.solver = 'snopt';
    setup.derivatives.supplier = 'sparseCD';
    setup.derivatives.derivativelevel = 'first';
    setup.method = 'RPMintegration';
    % Run GPOPS
    output = gpops2(setup);
    Solution = output.result.solution;
end

function [output] = Continuous(input)
    % This function references the full history components of the
    % optimal control problem.  It establishes the state derivatives
    % for the equations of motion as well as specifying the portion
    % of the Lagrange term in the cost function.

    global Scen
    mass      = Scen.m_sat + Scen.m_fuel;
    T_max     = Scen.T_max;
    mu        = Scen.mu;
    %% State and control Vector Inputs
```

```matlab
% EOM are computed in Equinoctal Elements    [a h k p q F]
a          = input.phase.state(:,1);
h          = input.phase.state(:,2);
k          = input.phase.state(:,3);
p          = input.phase.state(:,4);
q          = input.phase.state(:,5);
F          = input.phase.state(:,6);
A          = input.phase.control(:,1)*T_max/mass;
th         = input.phase.control(:,2);
psi        = input.phase.control(:,3);
%% Equations of Motion
% Equinoctal Reference Frame
n          = sqrt(mu./a.^3);
r          = a.*(1-k.*cos(F)-h.*sin(F));
G          = sqrt(1-h.^2-k.^2);
B          = 1./(1+G);
K          = 1+p.^2+q.^2;
x          = a.*((1-h.^2.*B).*cos(F)+h.*k.*B.*sin(F)-k);
y          = a.*(h.*k.*B.*cos(F)+(1-k.^2.*B).*sin(F)-h);
xd         = a.^2.*n./r.*(h.*k.*B.*cos(F)-(1-h.^2.*B).*sin(F));
yd         = a.^2.*n./r.*((1-k.^2.*B).*cos(F)-h.*k.*B.*sin(F));
% Partial Derivatives
dx_dk      = a.*(h.*B.*sin(F)-1);
dy_dk      = a.*(h.*B.*cos(F)-2.*k.*B.*sin(F));
dx_dh      = a.*(-2.*h.*B.*cos(F)+k.*B.*sin(F));
dy_dh      = a.*(k.*B.*cos(F)-1);
% Matrix Values
M11        = 2.*xd./(n.^2.*a);
M12        = 2.*yd./(n.^2.*a);
M13        = 0;
M21        = G./(n.*a.^2).*(dx_dk-h.*B.*xd./n);
M22        = G./(n.*a.^2).*(dy_dk-h.*B.*yd./n);
M23        = k.*(p.*x-q.*y)./(n.*a.^2.*G);
M31        = -G./(n.*a.^2).*(dx_dh+k.*B.*xd./n);
M32        = -G./(n.*a.^2).*(dy_dh+k.*B.*yd./n);
M33        = h.*(p.*x-q.*y)./(n.*a.^2.*G);
M41        = 0;
M42        = 0;
M43        = K.*y./(2.*n.*a.^2.*G);
M51        = 0;
M52        = 0;
M53        = K.*x./(2.*n.*a.^2.*G);
M61        = (G.*(h.*B.*dx_dh+k.*B.*dx_dk)-2.*x)./(n.*a.^2);
M62        = (G.*(h.*B.*dy_dh+k.*B.*dy_dk)-2.*y)./(n.*a.^2);
M63        = (q.*y-p.*x)./(n.*a.^2.*G);
% Disturbing Acceleration
```

```
    phi         = atan2(y,x);
    Ax          = A.*((sin(th).*cos(phi)-cos(th).*sin(phi))).*cos(psi);
    Ay          = A.*((cos(th).*cos(phi)+sin(th).*sin(phi))).*cos(psi);
    Az          = A.*sin(psi);
    % State Derivatives
    dynamics(:,1) =     M11.*Ax + M12.*Ay + M13.*Az;
    dynamics(:,2) =     M21.*Ax + M22.*Ay + M23.*Az;
    dynamics(:,3) =     M31.*Ax + M32.*Ay + M33.*Az;
    dynamics(:,4) =     M41.*Ax + M42.*Ay + M43.*Az;
    dynamics(:,5) =     M51.*Ax + M52.*Ay + M53.*Az;
    dynamics(:,6) = n + M61.*Ax + M62.*Ay + M63.*Az;
    %% Build output file
    output.dynamics  = dynamics;
    output.integrand = input.phase.control(:,1);
end

function [output] = Endpoint(input)
    % This function references the endpoint components of the
    % optimal control problem.  It establishes the terminal cost as
    % well as any applicable endpoint constraints (which are not
    % applicable to this problem).

    %% Define constants
    % Defined in Responsive_Orbits
    global Scen REF
    phi         = REF.phi;
    N_r         = REF.N;
    T_r         = REF.T;
    P           = REF.p;
    Q           = REF.q;
    %% Read relavent componants out of input structure
    a           = input.phase.finalstate(1);
    h           = input.phase.finalstate(2);
    k           = input.phase.finalstate(3);
    p           = input.phase.finalstate(4);
    q           = input.phase.finalstate(5);
    cf          = cos(input.phase.finalstate(6));
    sf          = sin(input.phase.finalstate(6));
    Lagrange    = input.phase.integral;
    %% Determine final state in the equinoctial reference frame
    % Misc quantities
    G           = sqrt(1-h.^2-k.^2);
    B           = 1./(1+G);
    % Position in ERF
    x           = a.*((1-h.^2.*B).*cf+h.*k.*B.*sf-k);
    y           = a.*(h.*k.*B.*cf+(1-k.^2.*B).*sf-h);
```

```matlab
    % Determine Rotation Matrix R_ir
    R_ir      = [ 1-P^2+Q^2    2*P*Q        2*P;
                  2*P*Q        1+P^2-Q^2    -2*Q;
                  -2*P         2*Q          1-P^2-Q^2]./(1+P^2+Q^2);
    % Determine Rotation Matrix R_im
    R_im      = [ 1-p^2+q^2    2*p*q        2*p;
                  2*p*q        1+p^2-q^2    -2*q;
                  -2*p         2*q          1-p^2-q^2]./(1+p^2+q^2);
    % Modify the maneuvering satellite into the reference satellite's orbital
    % frame coordinate system.
    zm_r      = R_ir*R_im'*[x;y;0];
    N_m       = zm_r(1)*cos(phi) + zm_r(2)*sin(phi);
    T_m       = zm_r(2)*cos(phi) - zm_r(1)*sin(phi);
    % Determine distance from reference satellite
    dT        = T_m-T_r;
    dN        = N_m-N_r;
    dz        = zm_r(3);
    %% Calculate cost
    if Scen.InPlane == 1
        ellipse     = (dT/100)^2 + (dN/10)^2 + (dz/10)^2;
    else
        ellipse     = (dz/10)^2;
    end
    Mayer           = 1/(1+exp(50*(ellipse-1)));
    output.objective = Mayer+Scen.alpha.*Lagrange;
end


function[ECI] = Convert_to_ECI(Output,Scen)
    % This function converts the thrust and angle solutions derived from
    % MATLAB into the Earth-Centered Inertial Reference frame.  It takes the
    % following inputs:
    %      [ECI] = Convert_to_ECI(solution,total,Scen)
    % Solution is a structure consisting of several fields listed below:
    %
    %      state:    The Equinoctal Elements for each time step
    %
    % Output is a structure consisting of several fields listed below:
    %
    %      T:        The thrust profile in Newtons for each time step
    %      theta:    The in plane angle in the equinoctal frame in radians
    %      psi:      The out of plane angle in the equinoctal frame in
    %                radians
    %
    % Scen is a structure consisting of several fields listed below:
    %
    %      m_sat:    The satellite mass in kg%
```

```
%
% The output ECI are the [x y z] components in the Earth-Centered Inertial
% Reference frame for each time step.

for count     = 1:length(Output.a)
    %% Read Output structure
    a         = Output.a(count);
    h         = Output.h(count);
    k         = Output.k(count);
    p         = Output.p(count);
    q         = Output.q(count);
    F         = Output.F(count);
    A         = Output.T(count)*1e-3/Scen.m_sat;                % km/sec^2
    th        = Output.theta(count);
    psi       = Output.psi(count);
    %% Calculate useful quantities to generate Equinoctial Frame vector
    %   and Rotation matrix
    cf        = cos(F);
    sf        = sin(F);
    G         = sqrt(1-h^2-k^2);
    B         = 1/(1+G);
    x         = a*((1-h^2*B)*cf+h*k*B*sf-k);
    y         = a*(h*k*B*cf+(1-k^2*B)*sf-h);
    phi       = atan2(y,x);
    %% Calulate Equinoctial Frame Acceleration Vector
    sth       = sin(th);
    cth       = cos(th);
    sph       = sin(phi);
    cph       = cos(phi);
    sps       = sin(psi);
    cps       = cos(psi);
    E(count,:) = [(sth*cph-cth*sph)*cps;
                  (cth*cph+sth*sph)*cps;
                  sps]*A;
    %% Calculate Rotation Matrix
    R         = [ 1-p^2+q^2    2*p*q       2*p;
                  2*p*q        1+p^2-q^2   -2*q;
                  -2*p         2*q         1-p^2-q^2]./(1+p^2+q^2);
    %% Caluclate ECI Acceleration Vector
    ECI(count,:)= R*E(count,:)';
    end
end

function[] = XLSWrite()
    % This function takes the output data from MATLAB and converts it
    % into an excel document for plotting.
```

```matlab
delete CurrentTestRun.xlsx;
%% Read GPOPS solution
t       = Out.t/60;                                             % hr
T       = Out.T;                                                % N
theta   = Out.theta*180/pi;                                     % deg
psi     = Out.psi*180/pi;                                       % deg
Out.r = Out.a.*(1-Out.e.^2)./(1+Out.e.*cos(Out.M));
count = 1;
while Out.T(count) == 0
count = count +1;
end
thrust_time = Out.t(count);
%% Read COE for ref and man from STK and interpret/concatenate
% [m.t,m.COE] = Elements(man, [0 Output.t(end)+15*60], 'C');
[m.t,m.COE] = STK.Elements(man, [0 Out.t(end)], 'C');
m.a         = m.COE(:,1);
m.e         = m.COE(:,2);
m.i         = m.COE(:,3);
m.omega     = m.COE(:,4);
m.w         = atand(tand(m.COE(:,5)));
m.M         = atand(tand(m.COE(:,6)));
m.lat       = m.COE(:,7);
m.nu        = m.COE(:,8);
m.t         = m.t./60;
count1 = 1;
while abs(thrust_time/60-m.t(count1)) ~= min(abs(thrust_time/60-m.t))
    count1 = count1 +1;
end
ref_angle   = m.lat(count1);
% [r.t,r.COE] = Elements(ref, [0 Output.t(end)+15*60], 'C');
[r.t,r.COE] = STK.Elements(ref, [0 Out.t(end)], 'C');
r.a         = r.COE(:,1);
r.e         = r.COE(:,2);
r.i         = r.COE(:,3);
r.omega     = r.COE(:,4);
r.w         = atand(tand(r.COE(:,5)));
r.M         = atand(tand(r.COE(:,6)));
r.lat       = r.COE(:,7);
r.nu        = r.COE(:,8);
r.t         = r.t./60;
%% Convert for 2-D plot
m.r         = m.a.*(1-m.e.^2)./(1+m.e.*cosd(m.nu));
r.r         = r.a.*(1-r.e.^2)./(1+r.e.*cosd(r.nu));
x           = 2.*m.r.*sind((r.lat-m.lat)./2);
y           = m.r - r.r;
```

```
        oop        = (m.i + m.omega - r.i - r.omega).*sind(m.lat - ref_angle);
        z          = 2.*m.r.*sind(oop./2);
        %% Generate ellipse values
        ellipse.x  = linspace(-100,100,1000);
        ellipse.y  = 10.*sqrt(1-ellipse.x.^2./100^2);
        circle.x   = linspace(-10,10,1000);
        circle.y   = 10.*sqrt(1-circle.x.^2./10^2);
        Filename = [cd '\CurrentTestRun.xlsx'];
        xlswrite(Filename,[Out.t./60 Out.T Out.theta.*180/pi Out.psi.*180/pi],1);
        xlswrite(Filename,[x y z],2);
        xlswrite(Filename,[Out.t./60 Out.r-6378 Out.e Out.i.*180/pi Out.omega.*180/pi...
                          Out.w.*180/pi Out.M.*180/pi],3);
        xlswrite(Filename,[r.t r.a-6378 r.e r.i r.omega r.w r.M],4);
        xlswrite(Filename,[m.t m.a-6378 m.e m.i m.omega m.w m.M],5);
    end

end
end
```

A.3 Systems Tool Kit® Function Library

```
classdef STK
% STK Library Explanation of Structure Fields
%
% Created by James Sales
%
% The structure 'Scen' was designed specifically for use in the STK
% library for my Thesis research but can be fairly easily adapted to work
% elsewhere.  Not all of the following fields are necessary for every
% function but this is a summary of all of the fields used in the library.
%
% Scen Structure Fields:
%       Centroid:       The Lattitude, Longitude, and Elevation of the
%                       desired ellipse for an Area Target.
%       COE:            The Initial State Classical Orbital Elements
%                       formatted as follows:
%                           [r_p   e   i   RAAN   w   nu]
%                       the Radius of Periapsis is in kilometers and all
%                       angles are in degrees.
%       ElevAngle:      Minimum Elevation Angle for Access to satellite.
%       EndTime:        The Scen end time formatted as follows:
%                           'DD MMM YYYY HH:MM:SS'
%       EngineName:     String for the desired engine name.
%       Epoch:          The Epoch time formatted as follows:
%                           'DD MMM YYYY HH:MM:SS'
```

```matlab
%       m_sat:          The satellite dry mass in kg.
%       m_fuel:         The fuel mass in kg.
%       Now:            Tracks time from Epoch to current maneuver in
%                       seconds.
%       Path:           The filepath for external file storage.
%       Size:           The semi-major axis, semi-minor axis, and bearing
%                       formatted as a vector for the desired ellipse for
%                       an Area Target.
%       StartTime:      The Scen start time formatted as follows:
%                           'DD MMM YYYY HH:MM:SS'
%       TimeStep:       Animation increment given in seconds.
%       Title:          A string describing the desired Scen title.
%                       This string must contain no spaces.
%       T_max:          Max thrust in kN for custom engine.
%
% For the function 'Out_to_text.m' an additional structure is used.
% The following fields are necessary for this function.
%
% Out Structure Fields:
%       length:         length of the time vector
%       t:              The time vector in seconds
%       ECI:            The Earth-Centered Inertial attitude vector
properties
end
methods(Static)
    function [uiapp, root] = Initialize(Scen)
        % This function initializes STK and passes back the applicable handles for
        % further use in MATLAB.  The function takes the following inputs:
        %
        %       [uiapp, root] = STK_init(Scen)

        %% Grab STK handle if already if running or open STK and retrieve handle
        %  if not running
        try
            uiapp = actxGetRunningServer('STK10.application');
        catch
            uiapp = actxserver('STK10.application');
        end
        root    = uiapp.Personality2;
        %% Close existing Scen and open a new one
        try
            root.CloseScen();
            root.NewScenario(Scen.Title);
        catch
            root.NewScenario(Scen.Title);
        end
```

```matlab
    %% Set Scen Preferences
    % Set Date/Time Format
    root.UnitPreferences.Item('DateFormat').SetCurrentUnit('UTCG');
    % Assign Scen time period
    scen                            = root.CurrentScen;
    scen.SetTimePeriod(Scen.StartTime,Scen.EndTime);
    scen.Animation.StartTime        = Scen.StartTime;
    scen.Epoch                      = Scen.StartTime;
    scen.Animation.AnimStepValue    = Scen.TimeStep;
    %% Set Animation to Start Time
    root.Rewind()
end

function [sat, MCS] = Astrogator(Name, root, Scen)
    % This function initializes a satellite in Astrogator and returns the
    % applicable handles for further use in MATLAB.  It takes the following
    % inputs:
    %
    %       [sat, MCS_root] = Astrogator(Name, root, Scen)

    %% Initialize Satellite
    scen                    = root.CurrentScen;
    missionStartDate        = scen.StartTime;
    sat                     = root.CurrentScen.Children.New(18, Name);
    sat.SetPropagatorType('ePropagatorAstrogator')
    sat.Graphics.Attributes.Intervals.RemoveAll;
    sat.Graphics.Attributes.Default.Inherit = 0;
    sat.Graphics.Attributes.Default.IsOrbitVisible = 0;
    % Create handle to the Astrogator portion of the satellite's object model
    prop                    = sat.Propagator;
    % Create handle to the MCS and remove all existing segments
    MCS                     = prop.MainSequence;
    MCS.RemoveAll;
    %% Define the Initial States
    % Create handle to the Initial States
    IS = MCS.Insert('eVASegmentTypeInitialState','Initial State','-');
    % Designate satellite and fuel masses
    IS.SpacecraftParameters.DryMass = Scen.m_sat;
    IS.FuelTank.FuelMass            = Scen.m_fuel;
    IS.FuelTank.MaximumFuelMass     = Scen.m_fuel;
    % Input orbital elements
    IS.SetElementType('eVAElementTypeModKeplerian');
    IS.Element.RadiusOfPeriapsis    = Scen.COE(1);
    IS.Element.Eccentricity         = Scen.COE(2);
    IS.Element.Inclination          = Scen.COE(3);
    IS.Element.RAAN                 = Scen.COE(4);
```

```matlab
            IS.Element.ArgOfPeriapsis      = Scen.COE(5);
            IS.Element.TrueAnomaly         = Scen.COE(6);
            % Sets the orbit Epoch for the mission start time
            IS.OrbitEpoch                  = missionStartDate;
        end

        function [Target] = Area_Target(Name, root, Scen)
            % This function initializes an Area Target in STK and returns the
            % applicable handles for further use in MATLAB. It takes the following
            % inputs:
            %
            %        Target = AreaTarget(Name, root, Scen)

            Size     = Scen.Size;
            Centroid = Scen.Centroid;
            Target   = root.CurrentScen.Children.New(2, Name);
            Target.AreaType = 'eEllipse';
            Target.AreaTypeData.SemiMajorAxis = Size(1);
            Target.AreaTypeData.SemiMinorAxis = Size(2);
            Target.AreaTypeData.Bearing       = Size(3);
            Target.Position.AssignGeodetic(Centroid(1),Centroid(2),Centroid(3));
            Target.AccessConstraints.AddNamedConstraint('ElevationAngle');
            Target.AccessConstraints.GetActiveNamedConstraint('ElevationAngle').Angle = Scen.ElevAngle;
        end

        function[Eng]=Create_Engine_Model(root, Name, T)
            % This function creates a custom engine model in the Componant Library and
            % returns the applicable handle for further use in MATLAB. It takes the
            % following inputs:
            %
            %        Eng = CreateEngingModel(root, Name, T)

            scen         = root.CurrentScen;
            EM           = scen.ComponentDirectory.GetComponents('eComponentAstrogator').GetFolder('Engine Models');
            ConstThrust  = EM.Item('Constant Thrust and Isp');
            ConstThrust.CloneObject;
            num          = EM.count;
            for count = 0:num-1
                if length(EM.Item(count).Name) > 23
                    if strcmp(EM.Item(count).Name(1:24),'Constant Thrust and Isp1')
                        Eng = EM.Item(count);
                    end
                end
            end
```

```
        Eng.Name    = Name;
        Eng.Thrust  = T;
    end


    function [prop] = Propagate(Name, t, MCS, Prop)
        % This function adds a propagation step to the given satellite in
        % Astrogator and returns the applicable handle for further use in MATLAB.
        % It takes inputs as follows:
        %
        %       [prop] = Propagate(Name, t, MCS, Prop)

        prop = MCS.Insert('eVASegmentTypePropagate',Name,'-');
        prop.PropagatorName = Prop;
        prop.StoppingConditions.Item('Duration').Properties.Trip = t;
    end


    function[AccessTimes] = Compute_Access(root, sat, target, clock)
        % This function takes two handles and computes coverage encounters over
        % the entire Scen.  However, the values it returns are specific to my
        % thesis work and will likely require modification for use elsewhere.  It
        % takes the following inputs:
        %
        %       [CoastTime, Duration] = ComputeAccess(root, sat, target, count)

        root.UnitPreferences.Item('DateFormat').SetCurrentUnit('EpSec');
        scen       = root.CurrentScen;
        access     = target.GetAccessToObject(sat);
        access.ComputeAccess;
        DP =access.DataProviders.Item('Access Data').Exec(scen.StartTime, scen.StopTime);
        Enter      = cell2mat(DP.DataSets.GetDataSetByName('Start Time').GetValues);
        Depart     = cell2mat(DP.DataSets.GetDataSetByName('Stop Time').GetValues);
        for count = 1:min(length(Enter),length(Depart))
            Entry(count,:) = RO.Time_Sequencer(clock, Enter(count));
            Exit(count,:)  = RO.Time_Sequencer(clock, Depart(count));
            Spaces(count,:) = '     ';
        end
        AccessTimes.DT    = [Entry Spaces Exit];
        AccessTimes.EpSec = [Enter Depart];
    end


    function[t_end]=Output_to_text(Scen, Out, L, count)
        % This function generates a text file conforming to the Astrogator *.a
        % thrust attitude external file input parameters.  It takes inputs as
        % follows:
        %
        %       t_end = Out_to_text(Scen, Out, L, count)
```

```
    Filename = [Scen.Path,Scen.Title,'Profile',num2str(count),'.a'];
    t        = Out.t+Scen.Now;
    t_end    = Out.t(end)-Scen.Coast(count)+Scen.Now;
    ECI      = Out.ECI;
    Epoch    = Scen.Epoch;
    Maneuver = [t ECI]';
    Points   = length(t)-L;
    Factor   = 20;
    Order    = 1;
    Body     = 'Earth';
    Axes     = 'Inertial';
    % Open file & begin writing data conforming to the STK format requirements.
    fclose('all');
    FID = fopen(Filename,'w');
    fprintf(FID,'stk.v.5.0\r\n \r\n');
    fprintf(FID,'BEGIN Attitude\r\n \r\n');
    fprintf(FID,'NumberOfAttitudePoints\t%1.0f\r\n',Points);
    fprintf(FID,['Scen Epoch\t\t',Epoch,'\r\n']);
    fprintf(FID,'Blocking Factor\t\t%2.0f\r\n',Factor);
    fprintf(FID,'InterpolationOrder\t%1.0f\r\n',Order);
    fprintf(FID,['CentralBody\t\t',Body,'\r\n']);
    fprintf(FID,['CoordinateAxes\t\t',Axes,'\r\n\r\n']);
    fprintf(FID,'AttitudeTimeECIVector\r\n\r\n');
    fprintf(FID,'\t%6.6f \t\t%8.8f \t\t%8.8f \t\t%8.8f \r\n',Maneuver(:,L+1:end));
    fprintf(FID,'\r\nEND Attitude');
    fclose('all');
end

function [M] = Maneuver_From_File(Name, MCS, Scen, index)
    % This function conducts a Finite Thrust Vectored manuever in Astrogator
    % and returns the applicable maneuver handle for further use in MATLAB.
    % It takes inputs as follows:
    %
    %      M = ITV_Maneuver(Name, MCS_root, Scen, index)

    Filename = [Scen.Path,Scen.Title,'Profile',num2str(index),'.a'];
    M = MCS.Insert('eVASegmentTypeManeuver',Name,'-');
    M.SetManeuverType('eVAManeuverTypeFinite');
    M.Maneuver.SetAttitudeControlType('eVAAttitudeControlFile');
    Att_Control = M.Maneuver.AttitudeControl;
    Att_Control.Filename = Filename;
    M.Maneuver.SetPropulsionMethod('eVAPropulsionMethodEngineModel',
Scen.EngineName);
    M.Maneuver.Propagator.StoppingConditions.Item('Duration').Properties.Trip =
Scen.Dur(index);
```

```matlab
        M.Maneuver.Propagator.PropagatorName = Scen.Prop;
    end

    function [M] = FTV_Maneuver(Name, MCS, v, t)
        % This function conducts a Finite Thrust Vectored manuever in Astrogator
        % and returns the applicable maneuver handle for further use in MATLAB.
        % It takes inputs as follows:
        %
        %       M = FTV_Maneuver(Name, MCS_root, Vector, Duration)
        global Scen
        M = MCS.Insert('eVASegmentTypeManeuver',Name,'-');
        M.SetManeuverType('eVAManeuverTypeFinite');
        M.Maneuver.SetAttitudeControlType('eVAAttitudeControlThrustVector');
        Att_Control = M.Maneuver.AttitudeControl;
        Att_Control.ThrustVector.AssignXYZ(v(1),v(2),v(3));
        M.Maneuver.SetPropulsionMethod('eVAPropulsionMethodEngineModel',
Scen.EngineName);
        M.Maneuver.Propagator.StoppingConditions.Item('Duration').Properties.Trip=t;
        M.Maneuver.Propagator.PropagatorName = Scen.Prop;
    end

    function [M] = ITV_Maneuver(Name, MCS_root, v)
        % This function conducts an Impulsive Thrust Vectored manuever in
        % Astrogator and returns the applicable maneuver handle for further use in
        % MATLAB. It takes inputs as follows:
        %
        %       M = ITV_Maneuver(Name, MCS_root, Vector)
        M = MCS_root.Insert('eVASegmentTypeManeuver',Name,'-');
        M.Maneuver.SetAttitudeControlType('eVAAttitudeControlThrustVector');
        Att_Control = M.Maneuver.AttitudeControl;
        Att_Control.DeltaVVector.AssignCartesian(v(1),v(2),v(3));
    end

    function [t, Elem] = Elements(sat, time, Type)
        % This function takes a satellite and returns its orbital element time
        % history. It takes the following inputs:
        %
        %       [t, Elem] = Elements(sat, time, Type)
        root = sat.root;
        root.UnitPreferences.SetCurrentUnit('DateFormat','EpSec');
        if Type == 'E'
            EE       = sat.DataProviders.Item('Equinoctial Elements');
            EEICRF   = EE.Group.Item('ICRF');
            EEResults = EEICRF.Exec(time(1), time(2), 5);
            t = cell2mat(EEResults.DataSets.GetDataSetByName('Time').GetValues());
```

```
            a = cell2mat(EEResults.DataSets.GetDataSetByName('Semi-Major
Axis').GetValues());
            h = cell2mat(EEResults.DataSets.GetDataSetByName('e *
sin(omegaBar)').GetValues());
            k = cell2mat(EEResults.DataSets.GetDataSetByName('e *
cos(omegaBar)').GetValues());
            p = cell2mat(EEResults.DataSets.GetDataSetByName('tan(i/2) *
sin(raan)').GetValues());
            q = cell2mat(EEResults.DataSets.GetDataSetByName('tan(i/2) *
cos(raan)').GetValues());
            F = cell2mat(EEResults.DataSets.GetDataSetByName('Mean Lon').GetValues());
            Elem = [a h k p q F];
        elseif Type == 'C'
            COE      = sat.DataProviders.Item('Classical Elements');
            COEICRF  = COE.Group.Item('ICRF');
            COEResults = COEICRF.Exec(time(1),time(2),5);
            t = cell2mat(COEResults.DataSets.GetDataSetByName('Time').GetValues());
            a = cell2mat(COEResults.DataSets.GetDataSetByName('Semi-major
Axis').GetValues());
            e =
cell2mat(COEResults.DataSets.GetDataSetByName('Eccentricity').GetValues());
            i =
cell2mat(COEResults.DataSets.GetDataSetByName('Inclination').GetValues());
            omega = cell2mat(COEResults.DataSets.GetDataSetByName('RAAN').GetValues());
            w = cell2mat(COEResults.DataSets.GetDataSetByName('Arg of
Perigee').GetValues());
            M = cell2mat(COEResults.DataSets.GetDataSetByName('Mean
Anomaly').GetValues());
            lat = cell2mat(COEResults.DataSets.GetDataSetByName('Arg of
Latitude').GetValues());
            nu = cell2mat(COEResults.DataSets.GetDataSetByName('True
Anomaly').GetValues());
            Elem = [a e i omega w M lat nu];
        else
            t = [];
            Elem = [];
            fprintf('Specified Type not recognized\n')
        end
    end
end
end
```

Appendix B. GPOPS-II Structure Architecture

GPOPS-II Data Structure

- GPOPS-II uses MATLAB's Structure Array to organize data
  - Element-by-Element organization
- Blocks in this presentation are color-coded
  - Differentiate structures from fields
- Unless otherwise noted, all structures/fields are required
- For optional user-specified structures, "(O)" will be appended to structures name
- (p) denotes an array of structures corresponding to phases used in problem
- (g) denotes an array of structures corresponding to event groups used in problem
- (m) denotes an array of structures corresponding to mesh iteration history

- Structure or a nested structure (structure within a structure)
  - Light blue box with thicker black border
- Field (data element) within a structure
  - Dark blue box with thinner black border
  - User-specified fields will have text underlined
  - GPOPS-II generated fields will not have text underlined
- Lower-level nested structure shown on subsequent chart for viewing ease
  - Light blue box with thickest border
  - Border color corresponding to boxes in subsequent chart with lower-level nested structure

GPOPS-II Data Structure

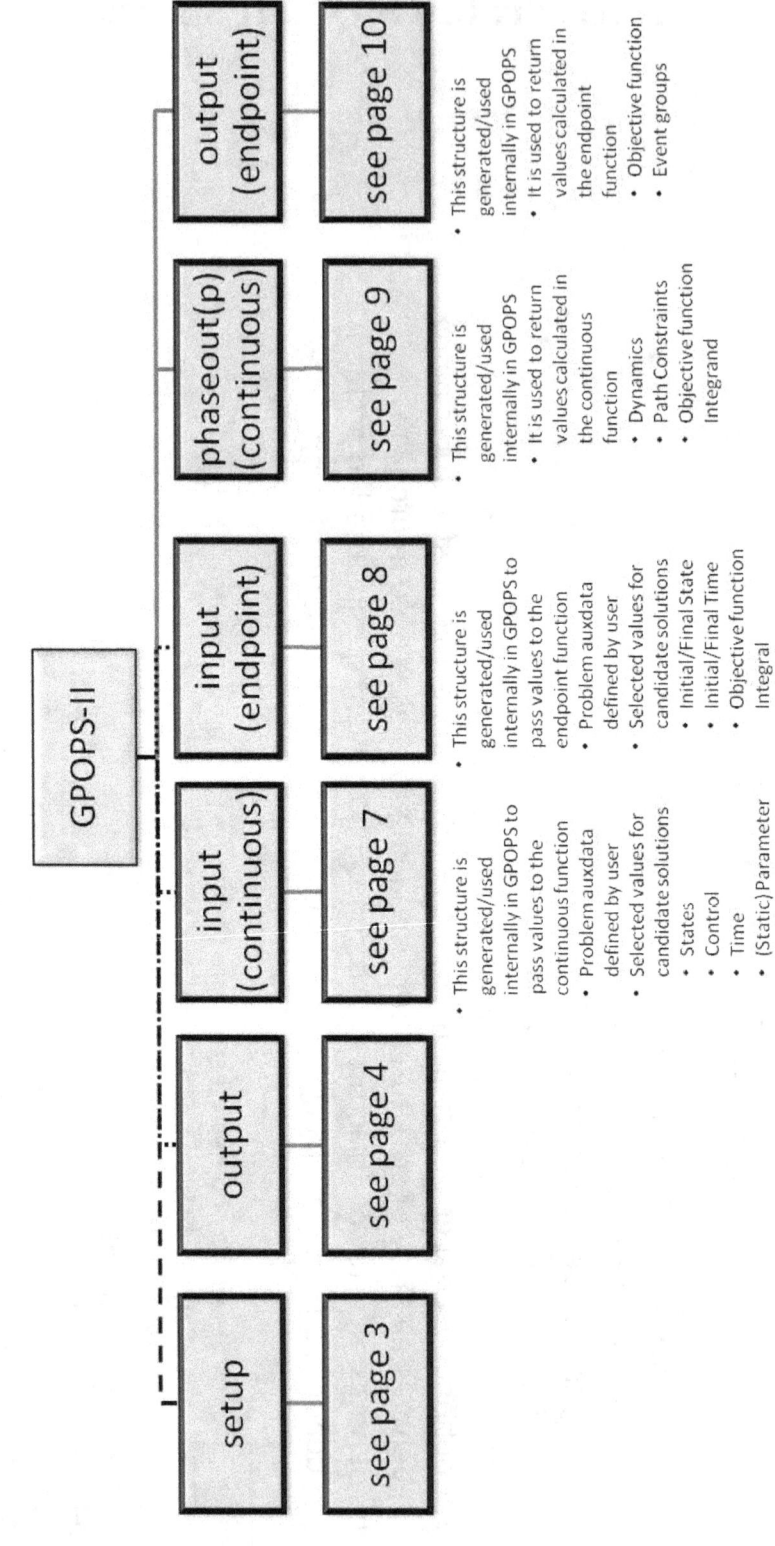

GPOPS-II Data Structure
setup Data Structure

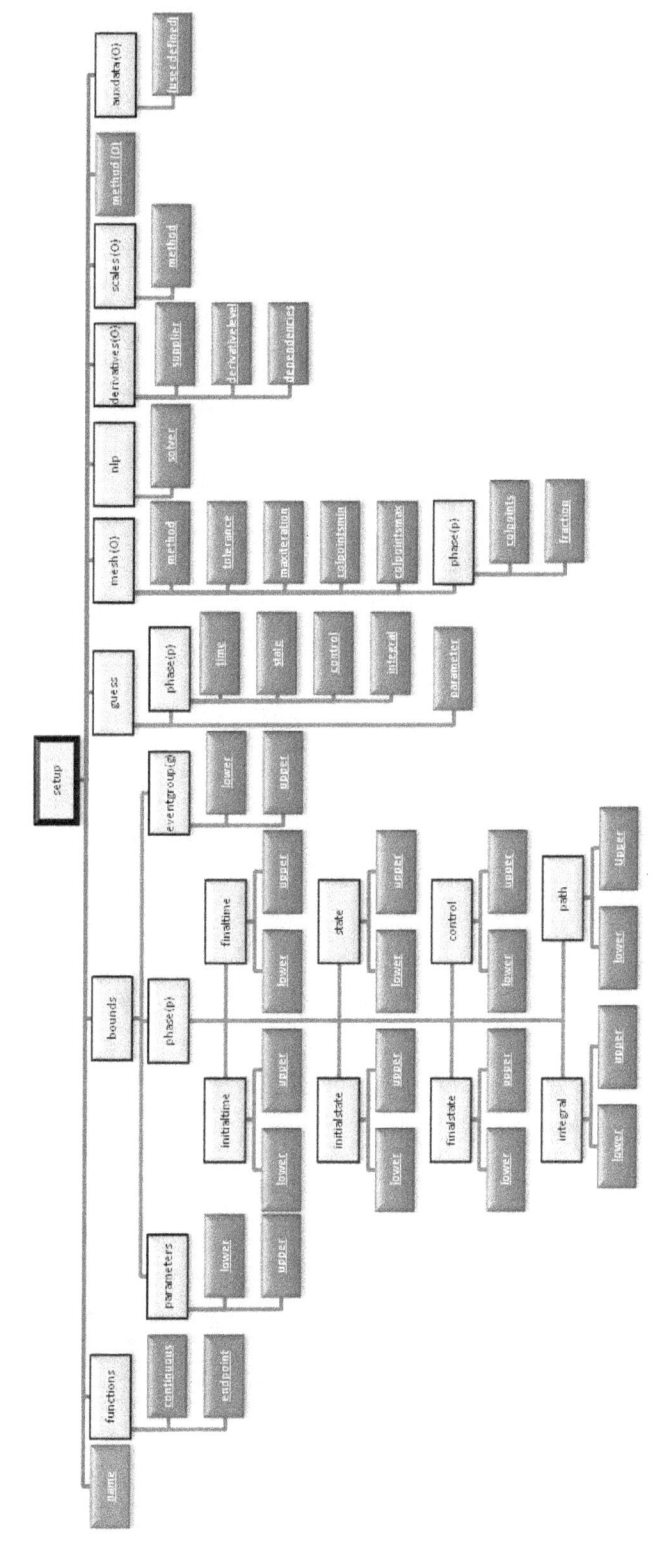

GPOPS-II Data Structure
output Data Structure

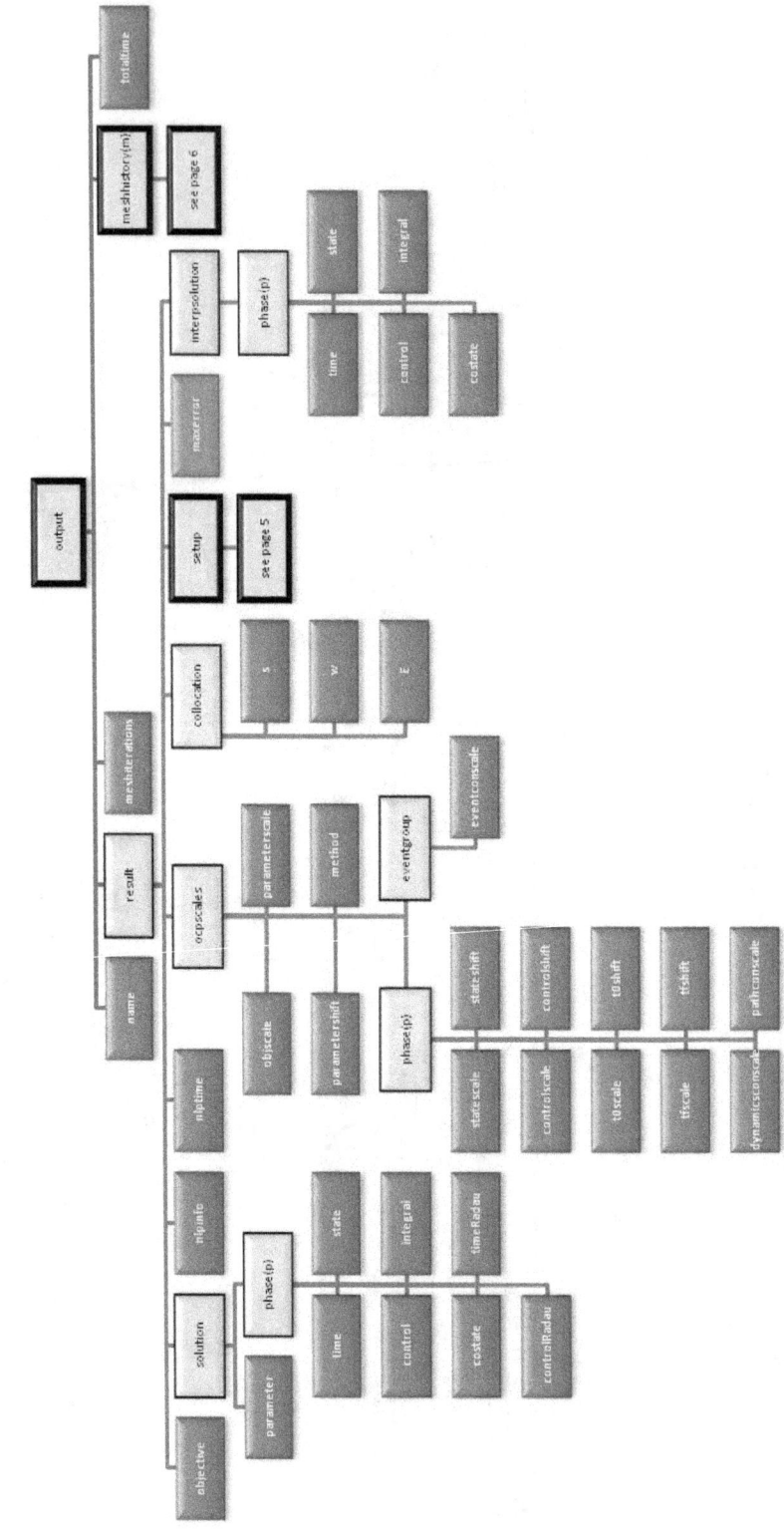

GPOPS-II Data Structure
output.result.setup Data Structure

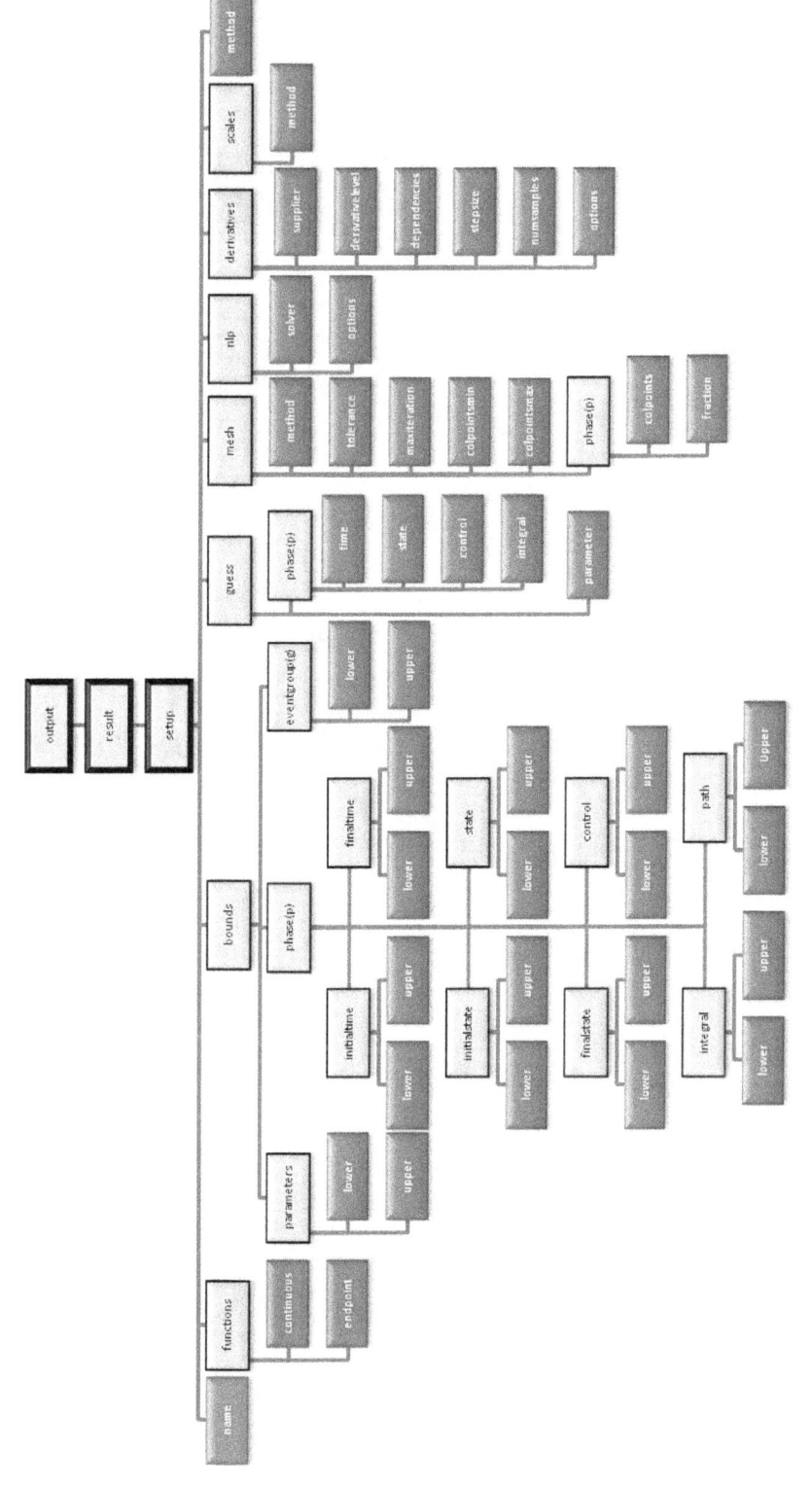

GPOPS-II Data Structure
output.meshhistory Data Structure

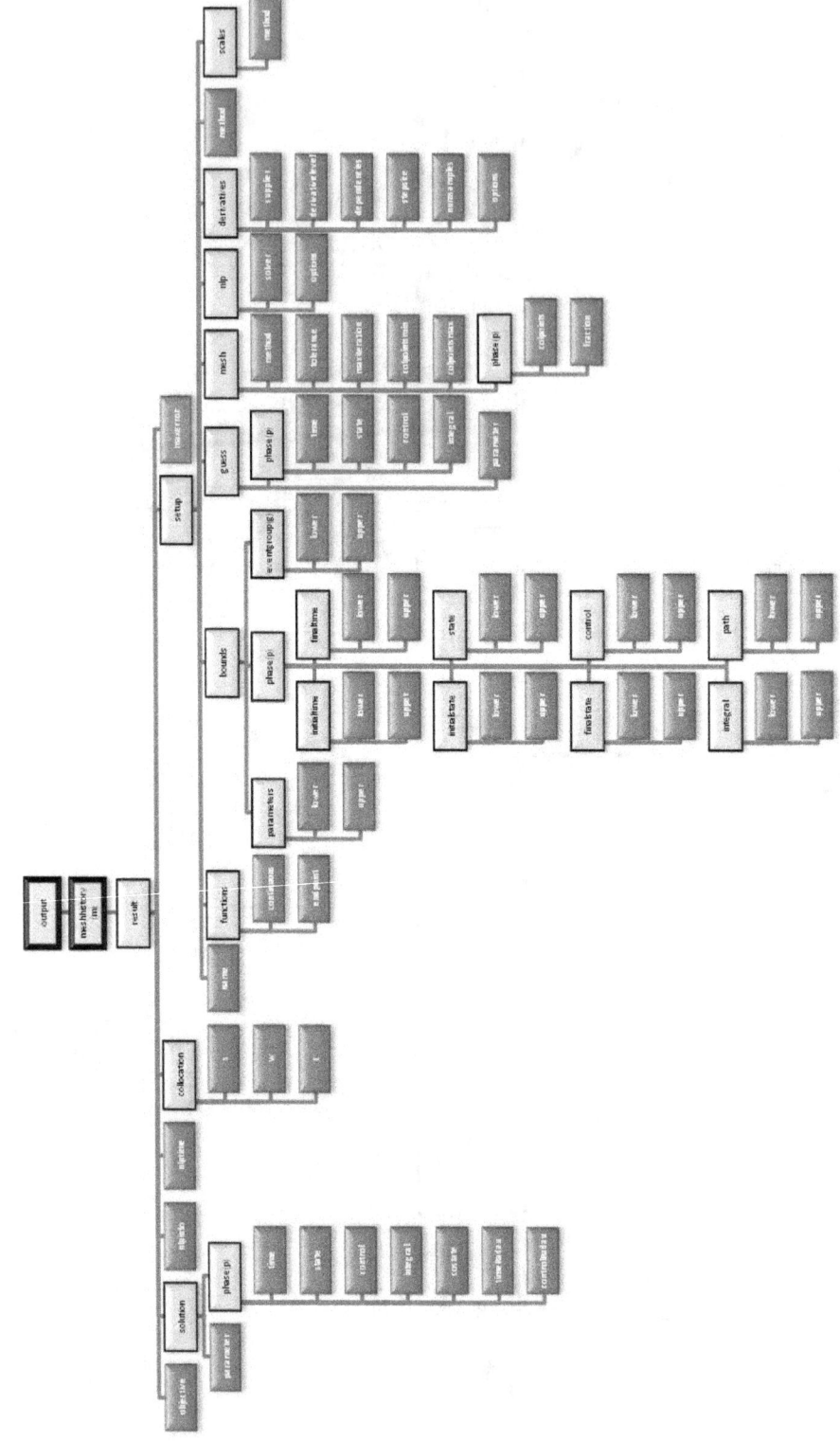

Any fields that were not included in the setup structure will be included with their default value

GPOPS-II Data Structure
input Data Structure (Continuous)

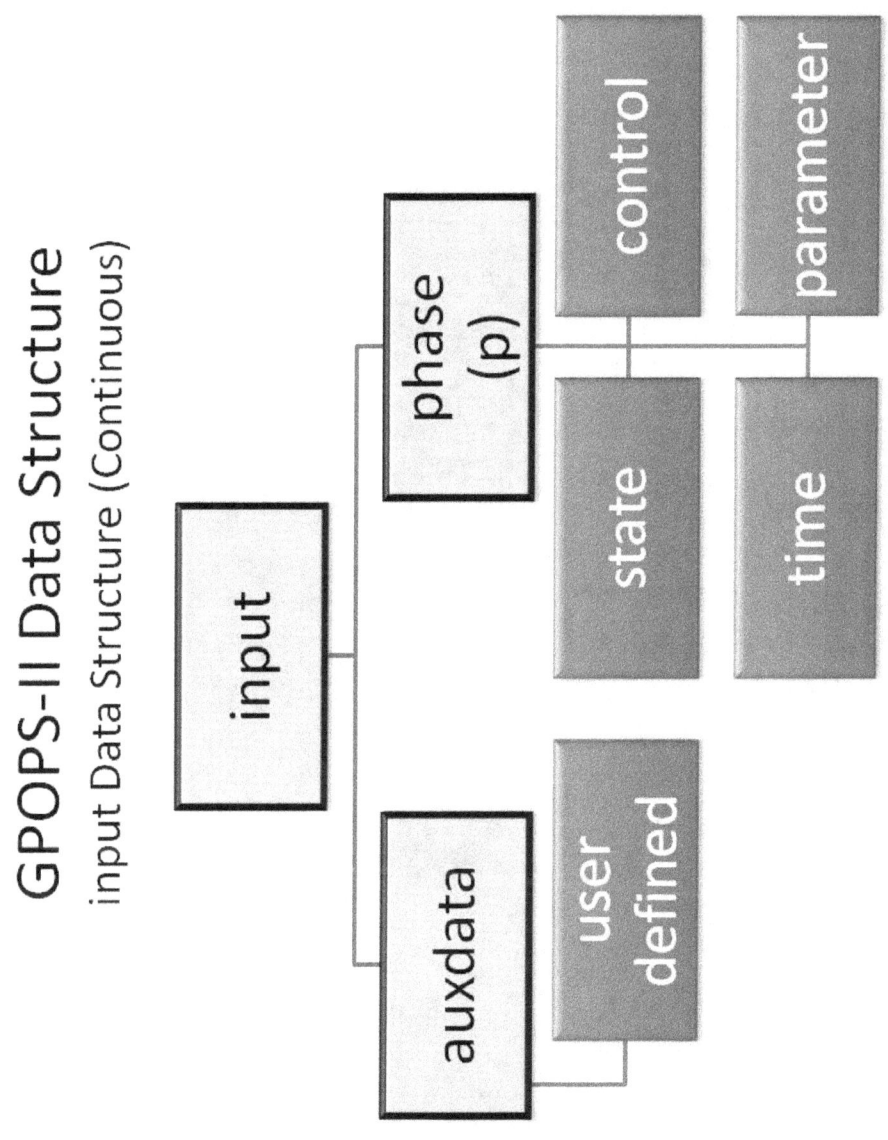

GPOPS-II Data Structure
input Data Structure (Endpoint)

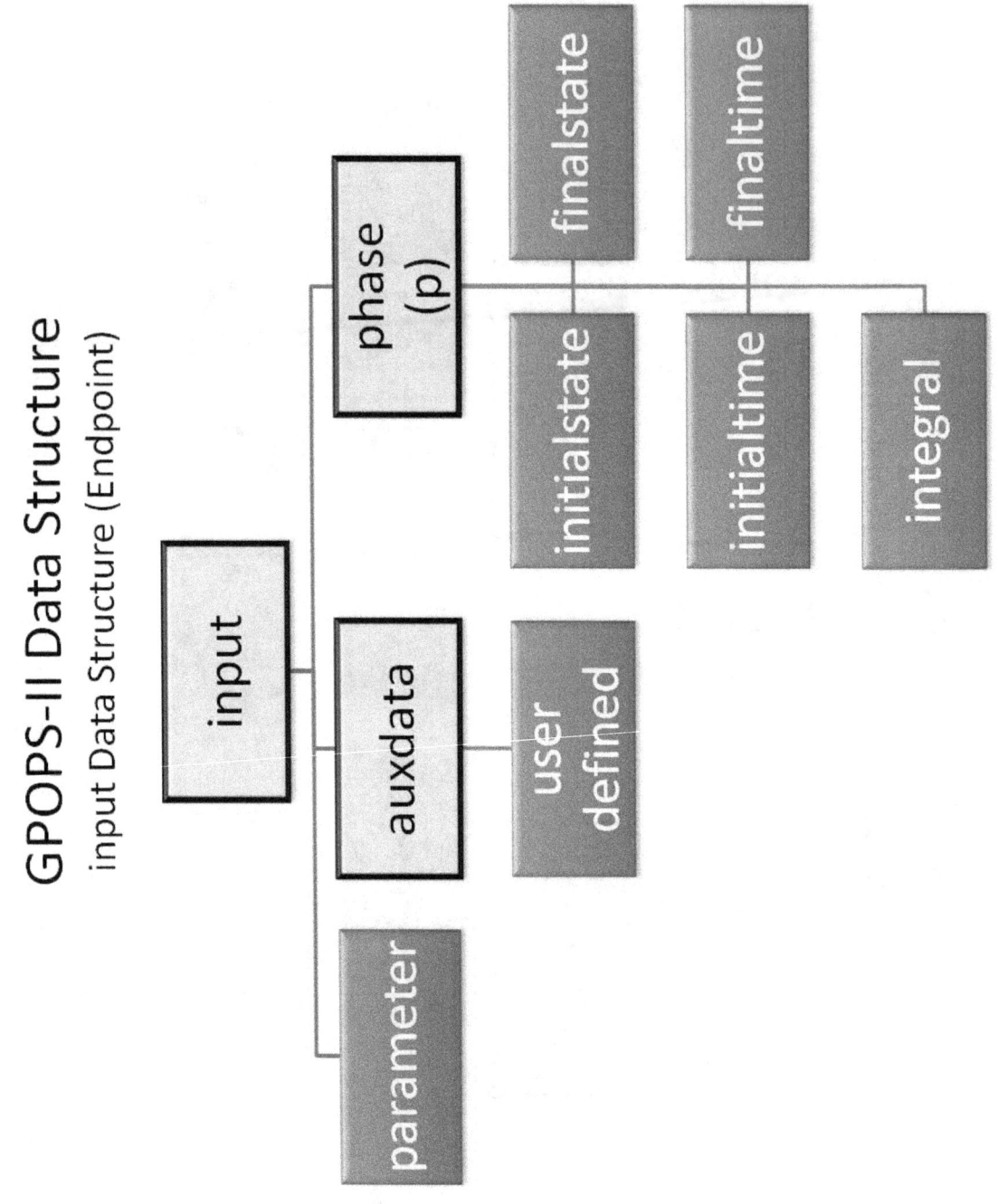

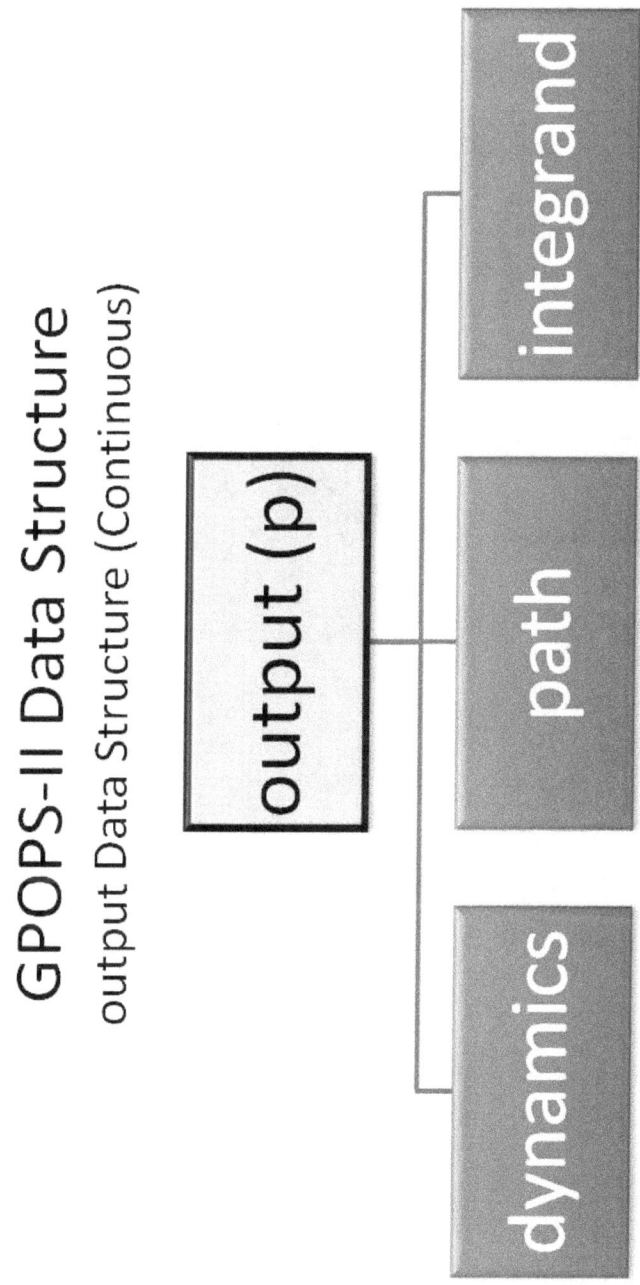

GPOPS-II Data Structure
output Data Structure (Endpoint)

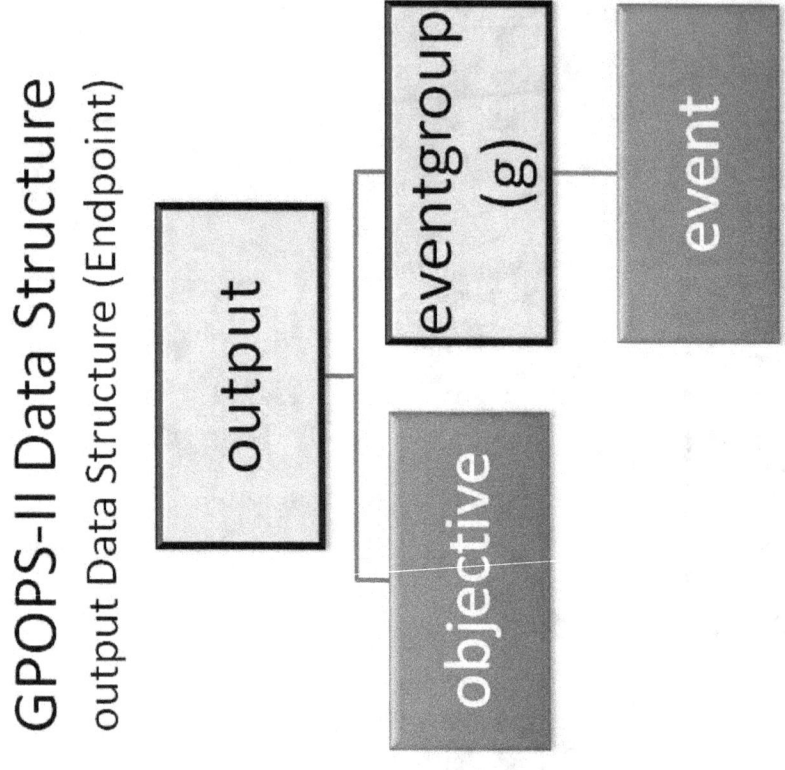

Bibliography

[1] J. C. Liou, *Active Debris Removal and the Challenges for Environment Remediation,* Houston: NASA Orbital Debris Program Office, NASA Johnson Space Center, 2012.

[2] "Instability of the present LEO satellite populatuons," *Advances in Space Research,* vol. 41, pp. 1046-1053, 2008.

[3] D. S. F. Portree, *Orbital Debris: A Chronology,* NASA-TP-1999-208856, (1999), 1999.

[4] W. E. Wiesel, Spaceflight Dynamics (Third Edition), Beavercreek: Aphelion Press, 2010.

[5] T. C. Co, *Operationally Responsive Spacecraft Using Electric Propulsion,* PhD Thesis, Graduate School of Engineering and Management, Air Force Institute of Technology, Wright Patterson AFB, OH, 2012.

[6] C. Zagaris, *Trajectory Control and Optimization for Responsive Spacecraft,* Master's Thesis, Graduate School of Engineering and Management, Air Force Institute of Technology, Wright Patterson AFB, OH, 2012.

[7] T. R. Jorris, *Common Aero Vehicle Autonomous Re-entry Trajectory Optimization Satisfying Waypoint and No-Fly Zone Constraints,* PhD Thesis, Graduate School of Engineering and Management, Air Force Institute of Technology, Wright Patterson AFB, OH, 2007.

[8] W. J. Karasz, *Optimal Re-entry Trajectory Terminal State Due to Variations in Wapoint Locations,* Master's Thesis, Graduate School of Engineering and Management, Air Force Institute of Technology, Wright Patterson AFB, OH, 2008.

[9] D. E. Yaple, *Simulation and Application of GPOPS for a Trajectory Optimization and Mission Planning Tool,* Master's Thesis, Graduate School of Engineering and Management, Air Force Institute of Technology, Wright Patterson AFB, OH, 2010.

[10] C. L. Darby, *hp-Pseudospectral Method for Solving Continuous-Time Nonlinear Optimal Control Problems,* PhD Thesis, University of Florida, Gainesville, FL, 2011.

[11] C. D. Hall and I. M. Ross, "Optimal Attitude Control for Coplanar Orbit Phasing Transfers," *Advances in Astronautical Sciences,* vol. 115, pp. 79-94, 2003.

[12] N. S. Bedrossian, S. Bhatt, W. Kang and I. M. Ross, "Zero-Propellant Maneuver Guidance," *IEEE Control Systems Magazine,* pp. 53-73, October 2009.

[13] D. A. Vallado, Fundamentals of Astrodynamics and Applications Third Edition, Hawthorne: Microcosm Press, 2007.

[14] W. E. Wiesel, Modern Astrodynamics (Second Edition), Beavercreek: Aphelion Press, 2010.

[15] H. Schaub and J. Junkins, Analytical Mechanics of Space Systems, Reston: American Istitute of Aeronautics and Astronautics, 2003.

[16] J. A. Kechichian, "Optimal Low-Thrust Rendezvous Using Equinoctial Orbit Elements," *Acta Astronautica,* vol. 38, no. 1, pp. 1-14, 1996.

[17] J. Kechichian, "Trajectory Optimization with a Modified Set of Equinoctial Orbit Elements. AAS/AIAA 91-524," in *Astrodynamics Specialist Conference,* Durango, CO, 1991.

[18] D. E. Kirk, Optimal Control Theory An Introduction, Mineola: Dover Publications, Inc, 1970.

[19] D. A. Benson, G. T. Huntington, T. P. Thorvaldsen and A. V. Rao, "Direct Trajectory Optimization and Costate Estimation via an Orthogonal Collocation Method," *Journal of Guidance, Control, and Dynamics,* vol. 29, no. 6, pp. 1435-1440, 2006.

[20] B. A. Conway, Spacecraft Trajectory Optimization, Cambridge: Cambridge University Press, 2010.

[21] A. V. Rao, D. A. Benson, C. L. Darby, M. A. Patterson, C. Francolin, I. Sanders and G. T. Huntington, "Algorithm 902: GPOPS, A MATLAB Software for Solving

Multiple-Phase Optimal Control Problems Using the Gauss Pseudospectral Method," *ACM Transactions on Mathematical Software,* vol. 37, no. 2, 2010.

[22] D. Garg, M. A. Patterson, C. L. Darby, C. Francolin, G. T. Huntington, W. W. Hager and A. V. Rao, "Direct Trajectory Optimization and Costate Estimation of Finite-Horizon and Infinite-Horizon Optimal Control Problems Using a Radau Pseudospectral Method," *Computational Optimization and Applications,* vol. 49, no. 2, pp. 335-358, 2011.

[23] D. Garg, M. A. Patterson, W. W. Hager, A. V. Rao, A. Benson and G. T. Huntington, "A Unified Framework for the Numerical Solution of Optimal Control Problems Using Pseudospectral Methods," *Automatica,* vol. 46, no. 11, pp. 1843-1851, 2010.

[24] D. Garg, W. W. Hager and A. V. Rao, "Pseudospectral Methods for Solving Infinite-Horizon Optimal Control Problems," *Automatica,* vol. 47, no. 4, pp. 829-837, 2011.

[25] A. V. Rao and M. A. Patterson, *GPOPS-II Version 1.0: A General-Purpose MATLAB Toolbox for Solving Optimal Control Problems Using the Radau Pseudospectral Method,* 2013.

[26] T. J. Masternak, *GPOPS-II Data Structure,* 2013, unpublished.

[27] Analytical Graphics Inc, *STK 10.0.1 Programming Interface,* 2013.

[28] J. R. Wright, *Orbit Determination Tool Kit Theory & Algorithms,* 2009.

[29] W. E. Wiesel, Modern Methods of Orbit Determination, 2nd ed, Beavercreek: Aphelion Press, 2010.

[30] Analytical Graphics, Inc, *ODTK A Technical Summary,* 2009.